LIFE SCIENCES FOR THE
NON-SCIENTIST

2nd Edition

LIFE SCIENCES FOR THE
NON-SCIENTIST

2nd Edition

Viqar Zaman

Formerly Head of Department of Microbiology,
National University of Singapore

 **World Scientific**

NEW JERSEY • LONDON • SINGAPORE • BEIJING • SHANGHAI • HONG KONG • TAIPEI • CHENNAI

Published by

World Scientific Publishing Co. Pte. Ltd.

5 Toh Tuck Link, Singapore 596224

USA office: 27 Warren Street, Suite 401-402, Hackensack, NJ 07601

UK office: 57 Shelton Street, Covent Garden, London WC2H 9HE

Library of Congress Cataloging-in-Publication Data
Zaman, Viqar.
    Life sciences for the non-scientist/Viqar Zaman.--2nd ed.
        p. cm.
    ISBN 981-256-282-6 (pbk. : alk. paper)
    1. Life sciences. I. Title.

    QH307.2. Z36 2005
    570--dc22

                                    2005041818

British Library Cataloguing-in-Publication Data
A catalogue record for this book is available from the British Library.

Disclaimer: The author of this book has attempted to trace the copyright holder of all the poems reproduced in this publication and apologizes to copyright holders if permission to publish in this form has not been obtained.

Printed in Singapore by Mainland Press

Dedicated to my DNA Carriers

Sara, Farah, Maryam, Sonia, Raisa, Ryan
and
Umar

Preface to the 2nd Edition

Most chapters in this edition have been revised and two new chapters on heredity and transplantation have been added. Each chapter is written as a stand-alone essay, giving readers the choice of selection without the necessity of reading any other section.

I have tried not to view science in isolation by including few poems and an occasional reference made to ancient mythology, which still affects our behavior and attitude till today.

We must not think that science has only provided material benefits to mankind. Science has been a powerful ally in the struggle against racism, social injustice and religious bigotry. It has drawn people away from superstition, quackery, witchcraft, black magic, demons and devils. It gives primacy to logic and reasoning. Above all, it has made us more respectful towards all forms of life in existance on the Earth. If this book is able to convince you of this message, I think my purpose of writing has been fulfilled.

VIQAR ZAMAN
MBBS, PhD, DSc, FRCP, FRC Path

Preface to the 1st Edition

Life science is a vast and ever expanding subject, and this book covers only a few selected topics of general interest in the field. The idea is to provide basic information on these topics without too many technical details.

The concept of liberal arts education originated in the United States, and has now been accepted by many schools and universities, including the National University of Singapore, where I have worked for many years. The liberal arts education ensures that those who enter a university, should be instructed not only in their chosen field, but also in the workings of nature, as revealed by science.

A scientifically illiterate society cannot expect to progress, and is bound to be relegated to the dustbin of history as has happened to many nations in the past. It was scientific and technical progress which enabled the European countries to colonize Asia, in spite of Asia being an older civilization. It is important that men of arts and humanities, jointly with the men of science, face the great ethical and moral dilemmas of modern times, and find a way to protect our environment and bring peace to this conflict ridden world.

This book is made up of essays accompanied with diagrams and photographs, to help in the understanding of the text. It has no references, but these could easily be obtained from the numerous text books on science, found in any library. I am hoping that the contents would be thought provoking and enjoyable, not only for the students, but also for any person who is simply interested in science. The scientist is a citizen of the world, and rises above the confines of nationality and race, in an effort to understand the great mystery of existence. By embracing science, one feels liberated from man-made ideologies and self-imposed shackles. This is a great feeling, which I wish to share with you through this book.

VIQAR ZAMAN
MBBS, PhD, DSc, FRCP, FRC Path

Acknowledgments

I thank my life-partner Rosy for reading the manuscript and making many helpful suggestions. To Drs. Mary Ng, P. Gopalakrishnakone and Ho Bow for the use of their electron-micrographs. To Aslam Bashir for the art work. To the technical staff of the departments of microbiology of the National University of Singapore and the Aga Khan University for their help over the years.

My thanks also to Dr. Rumina Hasan for the use of Microbiology and Pathology department facilities at the Aga Khan University; and to Mr. A. Raheem Essa for typing the manuscript.

I have borrowed thoughts and ideas from many sources and apologize if acknowledgement has not been made to all of them.

Finally, my thanks are due to Ms. Lim Sook Cheng and Linda Kwan of the World Scientific for their support and encourgement during the publication of this book.

Contents

1

In Search of Truth

The word science is derived from *Scientia*, which means knowledge in Latin. As knowledge is desired by all human beings, one would have expected science to be universally popular. Unfortunately, this is not the case because science has often undermined established beliefs, and therefore run into conflict with the orthodox elements of the society.

It is not possible to give credit to any single country for starting the scientific revolution, but the Greeks were probably the earliest contributors to scientific thought. The ancient Greeks had built a beautiful world of myths and legends, which explained virtually everything encountered by man. There were numerous gods, and they had to be continuously appeased, with prayers, incantations and sacrifice. But with the passage of time, philosophers arose, who started to express doubts about the validity of the ancient Greek beliefs. These doubters could be regarded as the earliest amongst the scientists: as without doubt, there is no questioning, and without questioning, there is no science. However, the doubters had a heavy price to pay, and the trial of Socrates is an example of this.

In his famous trial, Socrates (469–399 BC), was charged with impiety and corrupting of the youth. Socrates took his stand upon the abstract principles of his philosophy, according to which wisdom consists in knowing how little we know, and that the world can best be served by truth and virtue, through knowledge. As the trial went on, it became obvious that there was no desire on the part of the Athenians to impose the death sentence on Socrates, if he apologized or voluntarily left Athens. But Socrates did neither, and accepted the death sentence by drinking Hemlock, (a plant poison from the genus *Conium*). The death of this great man did not end the search for truth in Greece. Many great philosophers followed him, and eventually, it was the Greek philosophy which destroyed the myths and legends of the Greek religion. The Hippocratic School emerged, which did not look for divine influence in the causation of disease, but attributed it to environmental pollution and bad nutrition. This was the first secular concept of disease, which forms the basis of modern medicine.

Another conflict between science and religious orthodoxy occurred, when Copernicus published his book, *On the Revolution of Celestial Spheres*, in 1543. The Church banned it, as it proposed the Heliocentric or sun-centered concept of the solar system, whilst the church believed in the Geocentric or the earth-centered concept. Anyone who opposed the Church was severely punished; for example, in 1600, Bruno was burnt on the stake in Rome. In 1620, Vanini, who called himself a naturalist, was burnt on the stake in Toulouse. In 1621, Fountainier met a similar fate in Paris.

In 1663 followed the famous trial of Galileo because of his support and approval of the Copernican theory. For this he was brought before the inquisition, which pronounced him guilty of heresy and offered him absolution on condition of full abjuration. Unlike Socrates, Galileo submitted. He was asked to repudiate his belief that the earth goes around the sun, by declaring on his bended knee, that, "with sincere heart and unfeigned faith, I abjure, curse and detest the said error and heresy, contrary to the holy church, and I swear that I will never in future say or assert anything which may give rise to a similar suspicion..." Galileo was sentenced to house arrest for life and died a broken man in 1642. In 1992, the Holy See published an apology for the suffering inflicted on Galileo by the Church. This was a noble and a highly overdue gesture.

Galileo and Copernicus demonstrated a very important scientific concept, that our senses cannot always be relied upon to give correct information. The "rising" and "setting" of the sun is an illusion caused by the rotation of the earth. Neither do stars in the sky mysteriously "appear" during the night and "disappear" during the day. Similarly, the sky is not a solid dome, as the ancients believed, but an illusion, due to the absorption of light by the earth's gaseous atmosphere.

A major milestone of science was reached in the year 1859, when Charles Darwin published his famous book, *On the Origin of Species*. It hit the world like a bombshell, and again evoked a severe reaction from the Church. But the world had changed, and the Church was no longer as powerful as in Galileo's time. Darwin's ideas could not be suppressed. Evolution became a favorite topic of discussion all over the world. The "Darwinian revolution", as it was called, brought about a wholesale change even in the cultural values of the society. Humans were firmly established as members of the animal kingdom, and subject to all the natural laws under which life exists.

In principle, there is no absolute certainty in science. Science continually grows, expands and changes, in an effort to better understand the working of nature. It is not partial to any individual or doctrine. Its impartiality is its strength and although sometimes perceived as anti-religious, it is not. However, it does take issue with incorrect and misleading ideas and regards it as its duty to challenge and disprove them.

The scientific enquiry is generally but not always conducted by the so-called "deductive" approach, which is hypothesis driven. Hypothesis is the name given to the initial idea or ideas used to explain a certain phenomenon. This is followed by a theory, in which these ideas are collated and presented in an organized form. Peter Medawar (1915–1987), Nobel Laureate in Medicine (1960) compared the progress from hypothesis to theory, to that of a polymer built out of a monomer.

Theories are always tentative and a theory which was accepted in the past may be rejected in the future. A classic example of this is the replacement of Newton's theory of universal gravitation by Einstein's theory of relativity. To a non-scientist, this sounds like inconsistency but to a scientist this is the very basis of its success.

Scientific theories have a predictive value. For example, if the chemical structure of two substances is known, it should be possible to predict what will happen when they are mixed. An outstanding example of scientific prediction occurred in 1869, when a Russian scientist Dmitry Mendeleyev (1834–1907) correctly predicted that some day, elements unknown during his time, would be discovered to fit in the empty slots in the periodic table based on their atomic weights.

Theories exist in all disciplines, not just science. So how does one differentiate a scientific theory from a non-scientific one? Karl Popper (1902–1995), a famous philosopher of science, suggested a method for doing this. According to him, scientific theories are potentially falsifiable, while non-scientific ones are not. For example, the "big-bang" theory of the origin of universe is falsifiable and it is therefore science. In contrast, Freud's theory of psychoanalysis is not falsifiable and it is therefore "non-science". In scientific theories, there are at least some facts to grapple with; while non-scientific ones are purely imaginative, impossible to prove or disprove. However, this does not mean that theories which fall under "non-science" category are necessarily incorrect.

Karl Popper thinks that falsifiability is crucial for the advancement of science. It is the responsibility of scientists to try and falsify existing theories. By doing this, incorrect theories get eliminated, and this takes us closer to truth. He states in *Conjectures and Refutations* (1962) that, "The history of science, like the history of all human ideas, is a history of irresponsible dreams, of obstinacy, and of error. But science is one of the very few human activities, perhaps the only one, in which the errors are systematically criticized and fairly often, in time, corrected".

Thomas Kuhn (1922–1996), another eminent philosopher of science, points out that science advances by revolutionary changes or "paradigm shifts"; where one conceptual view gets replaced by another. This change does not always occur without resistance as "novelty only emerges with difficulty". This is because human nature dislikes change and scientists too prefer to work within an existing paradigm. A well-known example of paradigm shift occurred when Ptolemy's idea that the sun goes around the earth, was discarded in favor of Copernicus's view that the earth goes around the sun. This was a major paradigm shift, as it revolutionized the whole field of astronomy. There are numerous such examples in the history of science, showing that science never reaches a plateau when paradigm shifts cease to occur.

A number of scientific discoveries are a product of serendipity or sheer good luck. For example, Lithium discovered in 1817 was used originally for treating gout, but was observed to benefits patients who were mentally disturbed. It is now used as a mood stabilizing agent. Alexander Fleming (1881–1955) a Scottish bacteriologist serendipitously discovered Penicillin in 1928. He left cultures of a bacterium, *Staphylococcus aureus*, unattended for some time and when he re-examined the culture dish, he found it was contaminated by a mould which had inhibited the bacterial growth around it. The mould was *Penicillium notatum* producing the antibiotic penicillin which had seeped through the agar. In 1945, Alexander Fleming received the Nobel Prize for this discovery which marked the beginning of the antibiotic era.

Science places great importance on objectivity or unbiased observation. However, total objectivity is impossible, which explains why scientific experiments always have controls. For example, most drug trials are conducted using a "double-blind" method. Neither the doctor nor the patient knows which is the drug and which is the control or the inert substance until the trial is over. The code is then broken and the results statistically analyzed.

In science, well established ideas sometimes get categorized as "laws". To reach this status, they should have undergone numerous tests and trials. However, even the so-called "laws" are not as enduring as Homer's poetry or Beethoven's music. At this stage too there is no finality in science. The validity of the "law" is similar to the legal term "beyond any reasonable doubt". In practice, this approach works very well.

Scientific knowledge, from the moment it is created, belongs to, or should belong to, all of humanity. It is the fruit of human history and the collective effort of scientists, from all over the world. Justice and fairness should be practiced in the dissemination of scientific information. Unfortunately, national interests sometimes prevent this from happening, especially if the information is of military significance.

With the passage of time, science has become highly specialized. Scientific research is, therefore, mostly a team work. Hardly any major discoveries are published nowadays under a single authorship. This holistic approach is mainly responsible for innovation in science. Scientific fraternity is by and large a friendly one and international co-operation continued even during the cold war years between the USSR and the West. One hopes that this tendency will grow, as is being witnessed in the construction of the space station, and the proposed exploration of the solar system by Russia, the USA, Japan and other European nations.

Science flourishes in an atmosphere of intellectual freedom, where asking questions, and expressing views, carries no penalties. In societies where such freedom does not exist, science suffers and the societies remain backward.

Scientific research has been artificially divided into basic and applied. Amongst the industrial nations, Japan concentrated mainly on applied research, mostly to satisfy its consumer oriented industry. As a result, it has one of the largest number of industrial patents but relatively few Nobel Prize winners. Traditionally, pure or basic research has been done mostly in the universities. That this kind of research can ultimately turn out to be of practical value should not be underestimated.

Where do the radically new scientific ideas come from? Nobody knows, because we don't understand how the brain creates ideas. But why should they be any different from the inspiration that produces the great works of art or poetry? One can call it intuitive knowledge. For the religious minded, it becomes "divine inspiration", as Sir Ronald Ross, after discovering the mosquito as a vector of malaria, wrote:

> "This day relenting God
> Hath placed in my hand
> A wondrous thing..."

The structure of the Benzene ring came to Kekulé, a German chemist, in a dream of a snake chasing its own tail. So Kekulé wrote, "Let us learn to dream, gentlemen, then perhaps we may find the truth... but let us beware of publishing our dreams, before they have been put to the proof of the waking understanding".

There is a certain degree of nobility in science which is difficult to describe but has been eloquently expressed by a French mathematician and physicist, Henri Poincare (1854–1912). He states, "The scientist does not study nature because it is useful, he studies it because he delights in it, and he delights in it because it is beautiful. If nature was not beautiful it would not be worth knowing, and if nature were not worth knowing, life would not be worth living".

'Scientist Oath'

"I vow to strive to apply my professional skills only to projects which, after conscientious examinations, I believe to contribute to the goal of co-existence of all human beings in peace, human dignity and self-fulfillment.

I believe that this goal requires the provision of an adequate supply of the necessities of life (good food, air, water, clothing and housing, access to natural and man-made beauty), education, and opportunities to enable each person to work out for himself his life objectives and to develop creativeness and skill in the use of hands as well as head.

I vow to struggle through my work to minimize danger; noise; strain or invasion of privacy of the individual; pollution of earth, air or water; destruction of natural beauty, mineral resource and wildlife".

Meredith Thring
New Scientist, 1971

(The initial signatories of the oath included three Nobel Laureates — Sir John Kendrew, president of the International Council of Scientific Unions; Abdul Salam, director of the International Center for Theoretical Physics in Trieste; and Maurice Wilkins, co-discoverer of the DNA double helix)

✳✳✳✳✳

"Truth never dies. The ages come and go.
The mountains wear away, the stars retire.
Destruction lays earth's mighty cities low;
And empires, states and dynasties expire,
But caught and handed onward by the wise,
Truth never dies".

Anonymous
Untitled

2

From Molecules to Cell

Robert Hooke (1635–1703), a seventeenth century English physicist, first noticed "cells"[a] when he was examining a thin slice of cork under the lens of his home made magnifier. He described them as "Little boxes or cells distinct from one another". The term cell caught on and we still use it. Hooke did not realize the magnitude of his discovery and the revolution it would bring in the field of biology. There was a gap of almost two hundred years before the study of cell earnestly began and the credit for this goes mainly to a German physiologist by the name of Theodor Schwann (1810–1882), who first realized that all tissues, of both animals and plants, were made up entirely of cells. He wrote that "there is one universal principle of development for the elementary parts of all organisms, however different they may be, and that is the formation of cell". This comment, emphasizing the basic unity of life, went against the religious dogma of the uniqueness of man. It is said, that he first submitted his book about the cell, to the local bishop for clearance, in case he was accused of heresy! The book, after publication, created a great deal of excitement and many scientists became involved in the study of the cell. In 1858, another German physiologist, Rudolph Virchow (1821–1902), made his famous statement, "Omnis cellula e cellula" (all cells come from other cells), which relates to his

---

[a]Derived from Latin *cella*, meaning a cubicle.

finding that not the whole organism but only certain groups of cells become sick during an illness. He was the first to establish that cancer was a cellular disease.

Cells are broadly divided into prokaryotes and eukaryotes. All the bacteria are prokaryotes and the remaining cells are eukaryotes, which include the cells of multicellular animals and the unicellular protozoa. The main difference between the two types of eukaryotic cells is that the former are highly specialized, and perform only the function of the organ to which they belong. For example, the cells of the liver will perform only the function of the liver and the cells of the skin only that of the skin. Yet they work in unison in the interest of the animal. The animal body was therefore likened by Rudolph Virchow to a "cell state", with a social organization and an efficient work force. The disadvantage with this arrangement, however, is that the failure of only a few cells in any vital organ of the body can lead to the death of the whole organism. In contrast to this, the protozoan cell is self contained and capable of performing all the vital functions of life on its own. It appears that evolution sacrificed independence to interdependence, when complicated forms of life arose.

It is generally believed that the cell membrane (also known as the plasma membrane), came into existence first, before the other structures of the cell evolved. The cell membrane separates the cell from its environment and protects its contents. It is semipermeable, so that the excretory products can be expelled and nutrients drawn in. In other words, it is "leaky", but not too leaky, otherwise the cell contents would flow out.

The cell membrane has receptors on its surface which can detect specific chemicals. The receptors operate by a "lock and key" mechanism, i.e. only a certain key (molecule), will fit into a certain lock (receptor). In this manner, messages are sent from one part of the body to another. If the receptors are blocked due to disease or drugs, then the message would not be received, and organ failure occurs.

The cell membrane has lipids or fats as one of its main constituents. This gives it mechanical stability and flexibility. The flexibility of the membrane is best seen in moving amoeba or trophozoite, as it maintains its integrity, in spite of its constantly changing shape (Fig. 2.1). In amoeba and some other protozoa resistant stage known as cyst form, the cyst wall, made up of single or multiple layers (Fig. 2.2), protects the organism from adverse environmental conditions and ensures its survival for long periods. This is necessary for its transmission from one host to another.

Two nucleic acids (ribonucleic acid = RNA, and deoxyribonucleic acid = DNA) are found in all the cells, with the exception of viruses, which have either RNA or DNA. DNA contains the coded recipe for the manufacture of proteins. The messenger RNA (mRNA) conveys this information to an organelle called the ribosome, where the synthesis occurs. The flow of information from DNA to RNA to protein, just outlined, is regarded as the basis upon which all life on earth exists.

DNA is made up of a linear assembly of four different kinds of nucleotides. These nucleotides are designated by the initials of their constituent base: A (adenine), G (guanine), C (cytosine) and T (thymine). The *sequence* of nucleotides determines the genetic code which is unique for each species and each individual of the species. In a multi-cellular organism, every cell contains exactly the same DNA, with exactly the same order of bases. Why do cells then differ, becoming nerve cells, liver cells, muscle cells, etc? This occurs because different genes get activated in them. In other words, the genome is same in all the cells but the pattern of gene activity is different in different cells. This phenomenon is known as gene expression.

The shape of the DNA molecule was worked out by Watson and Crick in 1953 and consist of a pair of nucleotide chains twisted together to make the "double helix", which has been aptly called the "immortal coil", passed from generation to generation. The fundamental characteristic of DNA is that it is able to replicate and has been replicating since the beginning of life. As adults, we have trillions of cells in our body, but we all began as a single cell. It is a truly amazing accomplishment that at every division, our DNA has faithfully copied itself without much mistake.

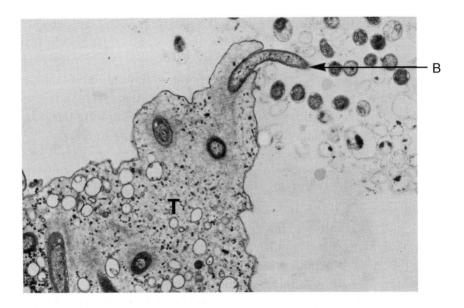

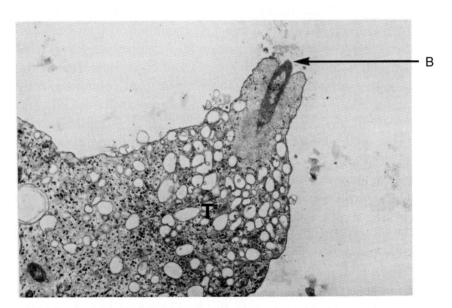

Fig. 2.1. *Entamoeba histolytica* trophozoites. Movement is directed towards food particles. In this case it is ingesting a bacterium (phagocytosis). During phagocytosis, pseudopods are thrown around the food particle, and it is drawn into the cytoplasm ×8000. TEM. B = bacterium, T = trophozoite.

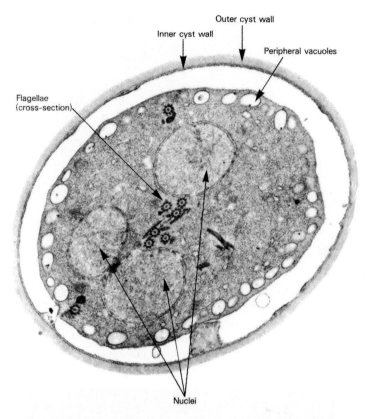

Fig. 2.2. *Giardia lamblia* cyst. The cyst wall has two layers. An outer thick layer and an inner thin layer which has double membrane. ×6000. TEM.

The nucleus contains 95% of DNA and is the brain or the control center of the cell. Inside the nucleus, the DNA is complexed with proteins to form a structure known as chromatin (Fig. 2.3). Before mitosis (cell division), the chromatin gets condensed into elongated structures called the chromosomes. Each animal and plant has a specific number of chromosomes. Humans have 23 pairs (46 in total). Each pair has two non-identical copies of chromosomes, which are inherited from each parent. The only cell without the nucleus is the erythrocyte or the red blood cell (Fig. 2.4). It is a biconcave doughnut shape disc with a thin membrane and contains hemoglobin, which gives it the red color.

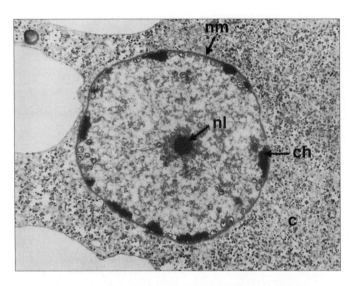

Fig. 2.3. The nucleus of the eukaryotic cell has a nuclear membrane attached to which are aggregations of chromatin. In the centre is a nucleolus. The cytoplasm of the cell has numerous granules which are ribosomal particles. This cell (*Entamoeba histolytica*) has no mitochondria. ×20 000. TEM. ch = chromatin, nl = nucleolus, nm = nuclear membrane, c = cytoplasm.

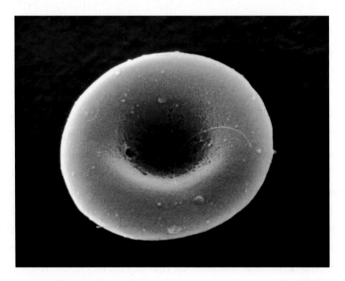

Fig. 2.4. Red cell or erythrocyte. It is a biconcave doughnut shape disc about 7.4 μm in diameter. ×5000. SEM.

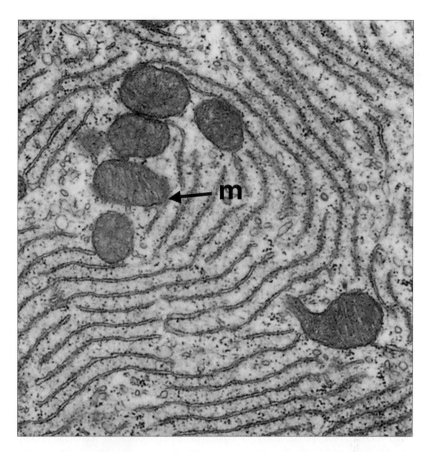

Fig. 2.5. Rough endoplasmic reticulum is covered with ribosomes which appear as minute dots. In the background mitochondria are also visible. ×26 000. TEM. m = mitochondria. (Courtesy of Mr. Yick Tuck Yong, Department of Anatomy, NUS)

The red blood cells do not multiply and only act as carriers of O_2 from lungs to tissues, and CO_2 from tissues to lungs. Their life span is approximately 120 days and the body needs to replenish their numbers continuously.

The cell contains membranous structures known as endoplasmic reticulum (ER), in its cytoplasm, which are of two types: rough and smooth. The rough ER is covered with ribosomes and the synthesized proteins pass through it (Fig. 2.5). One specialized

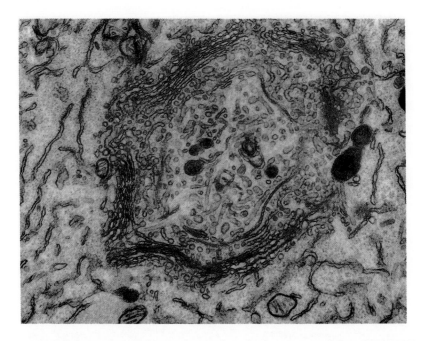

Fig. 2.6. Golgi complex made up of smooth endoplasmic reticulum. ×44 000. TEM. (Courtesy of Mr. Yick Tuck Yong, Department of Anatomy, NUS)

structure formed by the smooth endoplasmic reticulum is the Golgi complex (Fig. 2.6). Proteins synthesized in the rough endoplasmic reticulum migrate to the Golgi complex, where they get packaged into small vesicles. The vesicles are then transported to various sites where the proteins are needed. The cell cytoplasm also contains larger vesicle carrying enzymes, known as lysosomes, which are used for breaking down food and old organelles. They can be considered as the digestive system of the cell.

Another important structure in the cytoplasm of most eukaryotic cells are the mitochondria[b] (Fig. 2.7). This is essentially a transformed bacterium (protobacteria), that developed symbiotic relationship with the cell during the course of evolution,

---

[b]Derived from Greek *mitos* meaning thread and *chondrion* meaning granule.

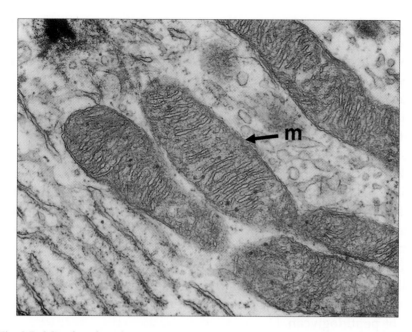

Fig. 2.7. Mitochondria showing shelves or cristae, made up of folded membranes. ×56000. TEM. m = mitochondria. (Courtesy of Mr. Yick Tuck Yong, Department of Anatomy, NUS)

approximately 1 billion years ago. It is composed of a complex membrane system, with the folded inner membrane forming shelves or cristae. It is on these shelves that the respiratory enzymes needed by the cell are located. Organisms which do not depend on O_2, may not have mitochondria, such as an amoeba, *Entamoeba histolytica*, and a flagellate, *Giardia lamblia*.

The number of mitochondria in the cell is related to its energy requirements. The muscle cells, for example, will have more mitochondria than the cells of the skin. Mitochondria have their own DNA and are capable of replication. Mitochondria are inherited through the ovum and not the sperm, therefore, all mitochondria are maternal in origin.

Cells have a protein scaffolding called the cytoskeleton. The cytoskeleton can be of three types — actin filaments, microtubules and intermediate filaments. Actin filaments are the most abundant and have a diameter of about 7 nm (nanometer). The microtubules are composed of a protein called tubulin and have a diameter of about 22 nm (Figs. 2.8(a), 2.8(b)). They can be clearly seen in cross-section of cilia and flagella (Fig. 2.9). The intermediate filaments have a diameter of about 10 nm. They usually line the interior of the nuclear membrane and extend outwards from the nucleus to the periphery of the cell.

In prokaryotes, the amount of DNA is much less than in eukaryotes and the genome of the prokaryote is an irregularly coiled naked DNA where the nuclear membrane is not present. Extrachromosomal DNA is often present in the form of a structure known as plasmid(s). They have the property of governing their own replication and are able to pass from one prokaryote to another by a process known as conjugation. Unlike the eukaryotic cell, the prokaryotes have no organelles in their cytoplasm. They have a rigid cell wall (with the exception of *Mycoplasma* sp) made up of peptidoglycan.

The usual method of cell division in prokaryote, and the single celled eukaryote (protozoa), is binary[c] fission (Fig. 2.10). The DNA molecule first replicates then attaches each copy to separate parts of the nuclear membrane in protozoa which then divides. The cytoplasmic division then follows, and the cell splits into two. The two daughter cells are of identical genetic composition. Under optimal conditions, a single bacterial cell usually divides every 20 minutes and with this rate of division, approximately five billion bacterial cells can be produced in 11 hours! In numbers, therefore, bacteria are the most prolific form of life on earth.

All the somatic cells in a multicellular organism divide by a complex process known as *mitosis*, during which each daughter cell receives identical number of chromosomes. If gametes were also to divide by mitosis, the union of sperm and ovum would double the number of chromosomes in each generation. This is prevented by a special kind of division, which ensures that the

---

[c]Derived from Latin *binarius* meaning paired.

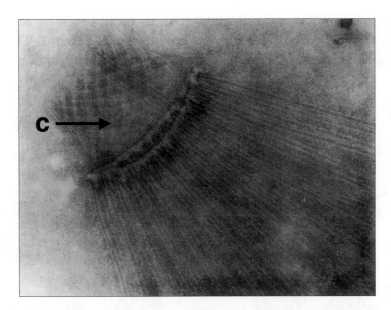

Fig. 2.8(a). Microtubules in a protozoan cell (*Toxoplasma gondii*). The anterior end, in which the microtubules criss-cross is known as the conoid, which is used for penetrating the host cell. ×20000. TEM. c = conoid.

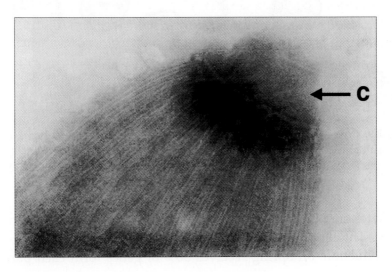

Fig. 2.8(b). Microtubules move the conoid — projecting it forward or drawing it in. In this case the conoid has been drawn in. ×20000. TEM. c = conoid.

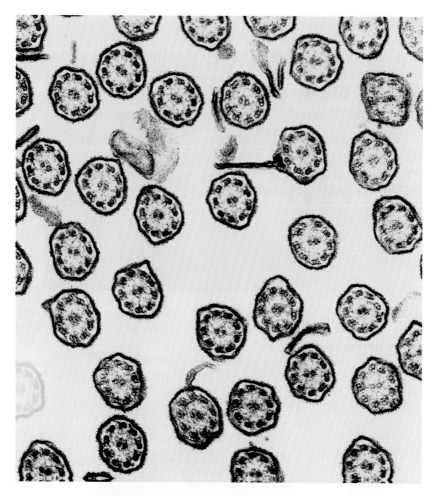

Fig. 2.9. Cross section of cilia showing 9 peripheral pairs of microtubules and a central pair. This structure is common to all the cilia wherever they are found. ×20000. TEM. (Courtesy of Mr. Yick Tuck Yong, Department of Anatomy, NUS)

gametes contain only half the number of chromosomes of somatic cells. This type of division is called *meiosis*.[d] Thus in sexual reproduction, off spring receives 50% of genes form each parent.

---

[d]Derived from Greek *meion* meaning to make smaller.

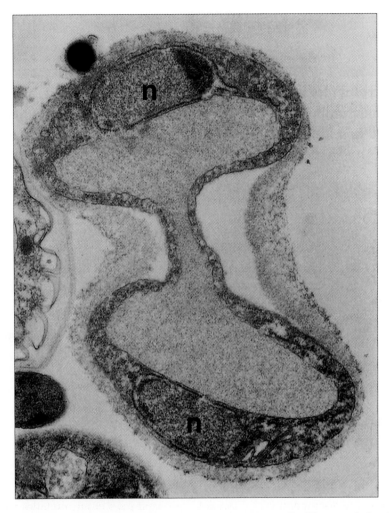

Fig. 2.10. A dividing cell, in this case a protozoan parasite *Blastocystis hominis*. The nuclear division occurs before cytoplasmic division and two oval nuclei are visible at opposite ends. ×24000. TEM. n = nucleus. (Courtesy of Ms. Josephine Howe, Department of Microbiology, NUS)

Viruses are generally regarded as non-living, as they are dependent on host cells for their survival and replication. They are, however, extremely successful as parasites, virtually infecting every living being, including bacteria.

Life's Criteria

A "living" organism has the following characteristics:

- The capacity to metabolise or produce energy.
- The capacity to divide or multiply.
- The capacity to respond or react to stimuli.
- The capacity to evolve or adapt to its environment.

✳✳✳✳✳

"To see a World in a Grain of Sand,
And a Heaven in a Wild Flower,
Hold infinity in the palm of your hand,
And eternity in an hour".

William Blake

3

From Inanimate to Animate

I have again and again grown like grass;
I have experienced seven hundred and seventy moulds.
I died from minerality and became vegetable;
And from vegetativenes I died and became animal,
I died from animality and became man.
Then why fear disappearance through death?
Next time I should die
Bringing forth wings and feathers like angels,
After that soaring higher than angels...
What you cannot imagine, I shall be that.

Jalaluddin Rumi
(In Idries Shah, *The Way of the Sufi*, New York: Dutton, 1970)

Jalaluddin Rumi was born in 1207, in Balkh, Northern Afghanistan, and followed the Sufic tradition of his father, Baha-uddin Mohammed. This is one of his most famous poems and was written approximately six hundred years before Darwin's theory of evolution was published.

The idea of evolution expressed by Rumi is probably the first of its kind, and contradicted the widely held religious belief that humans appeared on Earth in their present form. To Rumi, *change* is a fundamental attribute of nature for both the living and the non-living.

There are three main thoughts expressed in his poem. Firstly, that life arose from non-life ("minerality"). Secondly, that life moves from stage to stage, and presumably from less complex to more complex forms. Thirdly, that evolution does not cease at any juncture, and that there is a post-human future based on the continuing evolutionary process.

Clearly Rumi was not a scientist, neither did he base his idea on any scientific observations. These were purely subjective thoughts based on intuitive knowledge or imagination. In this context Rumi states, "When the psyche is in darkness one needs the light of reason to see his or her way through life, but when the psyche is illuminated one does not need reason's candle".

There are many theories about the *origin* of life on the Earth; however, for many scientists, evolution, which followed the origin of life is no longer a theory but a fact, which is discussed in Chapter 20.

In the 1970s, British astronomers, Fred Hoyle and Chandra Wickramasinghe proposed the theory of "panspermia" to explain the origin of life on the Earth. The "panspermia" theory postulates that life spreads from one heavenly body to another and that the Earth was seeded by primitive life forms in this manner. The seeding could occur during the passing of comets, by interplanetary dust or by meteorites. They further suggested that the seeding still continues and hence the advent of new viruses from time to time. In support of their theory, they were able to show that the inter-stellar dust had organic matter when it was examined spectroscopically.

In 1969, a meteorite fell in Murchison, Australia. Interestingly, the meteorite did show high concentrations of amino acids. This supported the idea of "panspermia". However, this theory still does not answer the question as to how life originated elsewhere in the universe, if indeed that is what happened.

Most scientists, however, think that life originated on the Earth from "non-life". To understand this process one has to go back to the cosmological events that resulted in the formation of the Earth. It is thought that our Sun came into existence some five billion years ago, out of the gas and debris left from an earlier supernova. A certain amount of this debris now orbits the sun as planets, including the Earth. The Earth, at the time of its birth, about 4.5 billion years ago, was very hot, due to radioactive decay and gravitational compression. It began to cool gradually and an atmosphere was formed by the emission of gases from volcanoes and rocks. The rocks themselves were rich in elements such as carbon, sulphur, silicon, phosphorus, oxygen and many metals, such as iron, calcium, potassium and sodium.

As the cooling progressed, the water vapour condensed to form the early rivers, lakes and seas. It is postulated that the water at this time was full of inorganic matter and contained a variety of molecules which were leached from the ground. These combined with the gases in the atmosphere to form the so-called "primordial soup", in which the amino acids, the precursors of proteins, formed after being subjected to powerful bolts of lightning, which were falling on the earth more or less continuously at that time.

Experimental evidence for this theory, which was first proposed by a Soviet biochemist Alexander Oparin, was produced in the early fifties by Stanley Miller, working in the department of Harold Urey, at the University of Chicago. He constructed a simple apparatus to simulate the conditions of the primitive earth. The apparatus was charged with methane (CH_4), ammonia (NH_3), hydrogen (H_2) and water vapor and subjected to electrical currents to replicate the effect of lightning. In the course of time, the water droplets that were collected from the lower outlet of the apparatus contained simple organic compounds and two amino acids, glycine and alanine. These are two of the twenty amino acids found in the proteins of all living organisms.

Oparin, in his book, *The Origin of life* (1924), explained why no new life forms are originating in the present day Earth. If by any chance a living organism approaching that of the first form of life should now appear, it would rapidly be destroyed by the O_2 of the air and be broken down by the countless microbes presently populating the soil and water. Therefore it is, impossible to witness the unique set of events that led to the origin of life billions of years ago.

Another group of scientists shifted their attention to the deep sea hydrothermal vents, which were first discovered by oceanographer Jack Corliss in 1977. These vents emit fluids containing dissolved metals, including sulphur, and the sulphur-metabolizing bacteria live in them. These bacteria act as the basis of the food chain for numerous other organisms including worms, clams and crabs. The formation of early life in this protected environment appeared a distinct possibility. Gunter Wachtershäuser, an organic chemist from the University of Regensberg, Germany, proposed a mechanism by which life could have originated from these ocean floor vents. In his concept, metabolism came before all else. Once a primitive metabolism was established, it began to run on its own and only after this did the cells basic elements, such as DNA and RNA, come into existence. The formation of acetic acid was the primary step with the catalytic metallic ion, iron sulfide, playing a key role. This was followed by the production of pyruvic acid, which reacted with ammonia to form amino acids. The amino acids, at high temperature, converted to form short protein-like chains called peptides, which were the forerunners of life.

Recently, attention has moved away from proteins, as the mother substance, to nucleic acids. Normally, proteins are formed according to the instructions coded in DNA, but DNA cannot replicate on its own without catalytic proteins, or enzymes. There is thus an impasse, as proteins cannot form without instructions from DNA and DNA cannot form without proteins. However, the situation is different with RNA. It was found by Thomas Cech and Sydney Altman of Yale University, for which they received the Nobel Prize in 1989, that certain types of RNA could act as their own enzymes. In other words, they can split themselves into two or replicate.

On the basis of this information, Walter Gilbert, a biologist at Harvard University, coined the term "RNA World", implying that life started as simple self-replicating RNA molecules. This

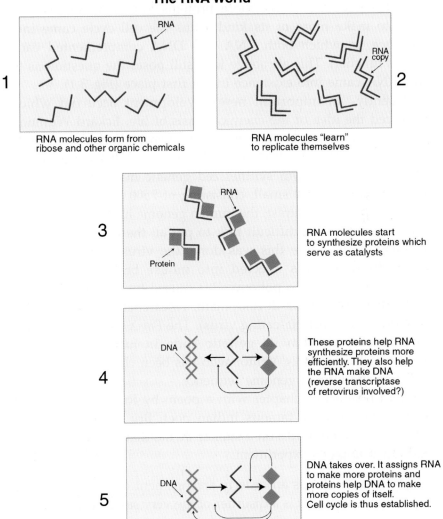

Fig. 3.1. According to this theory, RNA is the precursor of life. It passed through various phases of development to establish the present cell cycle. (Adapted from Scientific American, January, 1991). Original by Jason Kuffer.

eventually led to the formation of the double stranded DNA, and DNA took over as the main repository of genetic information. DNA used RNA to make more proteins, which in turn helped DNA to make more of its kind. Thus the cell cycle came into existence, in which both RNA and DNA complemented each other's function. This attractive idea still poses the question as to how RNA came into existence in the first place (Fig. 3.1).

Recently, an important new development occurred which advanced the idea of the chemical basis of life. Eckard Wimmer, professor of molecular genetics and microbiology at the State University of New York at Stony Brook, reported the creation of a live polio virus from synthesized genetic material. The polio virus genome is very small, consisting of 7500 chemical units or bases of RNA (by contrast, the human genome has 3 billion units), so it was not a very difficult task to put all the bases together in a proper order. Once this was done, the virus became "alive". When the virus was injected into mouse brain, it produced paralysis equivalent to poliomyelitis caused by the natural virus. In another spectacular development, researchers have created a bacteriophage (bacteria-eating virus). The *Phi-X174* bacteriophage was developed from its genetic code in just 14 days. The work was led by Craig Ventor who has been closely involved in mapping the human genome sequence.

I had begun this chapter with a poem by Rumi and will end with a couplet by a famous Indian poet, Brij Narain Chakbast (1882–1926), who writes on a similar theme that created the polio virus in Wimmer's experiment:

What is life? It is an orderly arrangement of elements,
What is death? It is a disorder of the very same elements.

✳✳✳✳✳

"Organic life beneath the shoreless waves
Was born, and nurs'd in ocean's pearly caves;
First forms minute, unseen by spheric glass,
Move on the mud, or pierce the watery mass;
These as successive generations bloom,
New powers acquire, and larger limbs assume;
Whence countless groups of vegetation spring,
And breathing realms of fin, and feet, and wing".

Erasmus Darwin
(Grandfather of Charles Darwin)

4

The Mother of All Cells

On an average, 300 million sperms are ejaculated by the male human during each act of coitus. Only a few hundred of these are able to reach the fallopian tube (tube connecting the ovary to the uterus), where the ovum or the egg is located. The sperms actively swim around the ovum and try to penetrate a layer of cells around it, known as the zona pellucida. The zona pellucida can recognize sperms of foreign species and prevent their entry unless they are closely related. However, if the zona pellucida is removed, cross fertilization is possible, but the hybrid ovum does not survive. For example, if the zona pellucida of the hamster ovum is removed, human sperms are able to penetrate it, but no future development follows. In fact, this procedure is used to test sperm vitality in patients with fertility problems.

As the sperms jostle around the ovum, only one succeeds in entering it. This prevents polyspermy, and is accomplished by the rapid depolarization of the ovum's plasma membrane, subsequent to the sperm's entry. Once fertilization has occurred, the process of embryogenesis begins. In 30 hours the cell cleaves and two cells appear. This process continues until a 16 cell stage, known as the morula, is reached. The morula develops a cavity and it is then called a blastocyst. The mother cells, which will make the embryo, appear on one side of the blastocyst in the form of a small mass or a clump. This clump of cells is known as the inner cell mass, and these cells are pluripotent stem cells, i.e. are able to generate many other cell types that form the body.

Fertilization and Cleavage of the Zygote

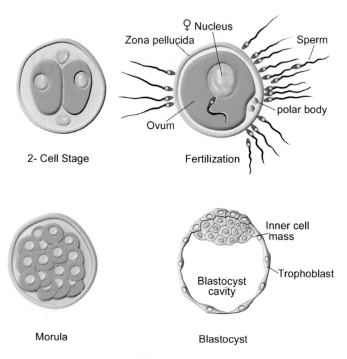

Fig. 4.1. Fertilization occurs by the entry of a single sperm into the ovum. The fertilized ovum is known as the zygote. The progressive division of the zygote leads to the formation of the morula (16-cell stage) and then the blastocyst. The cells of the inner cell mass are used as stem cells as they are pluripotent.

However, the inner cell mass on its own cannot develop into a fetus because the placenta is formed by the outer, or the trophoblastic layer of the blastocyst (Fig. 4.1).

Though the existence of the stem cells has been known for a long time, their therapeutic potential has only been discovered in the recent past. The researchers at Johns Hopkins and the University of Wisconsin, USA, pioneered these studies. For this work to proceed, two main problems had to be solved. The first was to develop and grow stem cells *in vitro* (outside the body). For this, the stem cells needed to grow atop embryonic mouse cells, known as the "feeder cells". The mouse cells provided nutrition to the stem cells but this close association raised the problem of safety, as the mouse cells could carry unknown viruses. Recently, Ariff Bongso and his team of researchers from the National University of Singapore showed that feeder cells can be obtained from human sources, thereby obviating the danger associated with the use of mouse cells.

The next difficult task was to find a method of inducing the stem cells to develop into the required specialized cells. Although much still needs to be done, by and large, these problems have been overcome. In order to assess if the laboratory grown stem cells actually repaired the tissues, the team at Johns Hopkins tampered with the spinal cords of laboratory bred mice, by infecting them with a virus which causes paralysis. When the stem cells were injected into the diseased mice, they regained most of their activity. In a similar experiment carried out by scientists form Israel and Spain, the stem cells could be induced to produce insulin in the pancreas of diabetic mice.

The stem cells do not entirely disappear from the body on its reaching maturity. Certain tissues of the body require constant replenishment; these include the skin, the intestinal epithelium and the blood cells. However, it was thought that this happens through "determined" cells, which can generate only their kind. It has recently been shown that this is not the case. Scientists from the University of Minnesota found that some stem cells from the bone marrow, are pluripotent, i.e. capable of forming more than one tissue. The advantage of using adult stem cells over the embryonic stem cells is that the former could be obtained and injected back into the same individual, thereby avoiding rejection. Rejection occurs when foreign cells are used. In spite of this discovery, it is felt that research should continue with both the embryonic and adult stem cells, as their relative advantages and disadvantages are not yet fully known.

It is estimated that some 128 new human embryonic cell lines have been produced worldwide at the time of writing this article. Some cell lines have been produced from embryos carrying genetic diseases to enable a study of these diseases, as well as their possible treatment.

It has been recently suggested that the length of our lives may be determined by the health of our stem cells. According to this theory, we age because our stem cells weaken and can no longer produce a fresh supply of cells to keep our organs young. If this is the case, then injecting new stem cells could extend the life span and delay the onset of old age. This is an area of active and potentially very fruitful research.

According to biologist Evan Snyder, Burnham Institute in LaJolla, California, stem cells could be used to detect and destroy cancer cells. He modified stem cells taken from a human embryo and added a gene that made the cells express an antitumor molecule known as TRAIL. When these cells were injected into mice with brain tumor, they homed into the tumor and released TRAIL, reducing the tumor size by as much as 70%. Snyder thinks that stem cells could be modified to express other anticancer molecules and thus become useful in cancer therapy.

In short, the age of "cell therapy" has arrived, and it may prove to be a great revolution in the history of medicine. We can look forward to the treatment of many disabling and degenerative diseases including Diabetes, Alzheimer's disease, Parkinson's disease, spinal cord injuries, retinal degeneration, heart diseases, stroke, burns, arthritis and the like. In fact, in all circumstances where tissue damage has occurred and the replacement of cells is needed, they would benefit from this therapy. The bone marrow transplantation has already proven that hematopoietic stem cells can be used for the treatment of blood cancers, congenital blood disorders and in diseases of bone marrow failure, such as aplastic anemia.

Stem Cell Sources

1. **Frozen human embryos**
   During the *in vitro* fertilization programme offered by scores of fertility clinics around the world, women are given hormones to produce eggs. This results in the discharge of multiple ova of which only one is implanted in the uterus. The rest are frozen and stored in clinics to be eventually discarded, if they are no longer required by the donors. It is estimated that in the US alone there are approximately 100 000 frozen fertilized eggs which could be used for stem-cell research. This is the easiest and, probably the best source of stem cells.

2. **Embryos created specifically to obtain stem cells**
   While the procedure for this remains the same as the one for frozen embryos, the purpose is different. For some, this method is highly objectionable and tantamount to infanticide as the embryo is being created solely for supplying the stem cells to certain individuals.

3. **Stem cells obtained from the patient's own cells**
   It is a known fact that some cells found in the skin and blood can be used for this purpose. In this technique, the nucleus of the donated ovum is first removed and the nucleus from the patient's skin or blood cell is injected in its place. This technique is similar to that used for cloning "Dolly", the sheep. The main advantage in this method is that the embryonic cells produced carry the patient's DNA, and are less likely to be rejected by the host.

4. **Cord blood**
   Soon after delivery, blood from the umbilical cord can be easily collected. Cord blood contains a large number of stem cells. After collection they can be frozen for future use. Also, no ethical issues are involved in this procedure.

5. **Bone-marrow**
   All blood cells and their precursors are formed in the bone-marrow. The progenitors for erythropoiesis (red-cell generation), granulopoiesis (white-cell generation) and megakaryopoiesis (platelet generation) occur in the bone-marrow. It now appears that some pluripotent stem cells are also found in the bone marrow, which are capable of generating cells other then that of blood.

✳✳✳✳✳

"All things in nature work silently. They come into being and possess nothing. They fulfill their function and make no claims. All things alike do their work, and then we see them subside. When they have reached their bloom each returns to its origin. Returning to their origin means fulfillment of their destiny. This reversion is an eternal law. To know that law is wisdom".

Lao-tze
Sayings of 6th Century BC

5

Cells Out of Control

All organs of the body are made up of a mixture of cell types, each of which performs a different function but in harmony with one another. They have a fixed location and do not move around like the wandering macrophages or the white blood cells. The retention of their form, function and growth is under the control of their genes. Two types of genes are involved, proto-oncogenes, which encourage growth, and tumor suppressor genes, which inhibit growth. In a normal situation, these are in balance with each other, ensuring an orderly pattern of growth. Each cell continuously senses its environment, by receiving signals from its immediate neighbors, and reproduces only when instructed by the cells in its vicinity. In this way territorial integrity is maintained and undue proliferation is avoided. Both these factors are violated by the cancer cells. Firstly, the cancer cells proliferate without any restraint and secondly, they invade territories that do not belong to them.

The cell–cell adhesion molecules, which keep cells in their proper location, appear to be absent in cancer cells, thus accounting for their spread. The spread could occur via the lymphatics or blood, leading to metastases[a] or dissemination of cancer to various parts of the body. It is this factor which makes cancer a dangerous and a life-threatening disease (Fig. 5.1).

---

[a]From the Greek meaning "change in location".

37

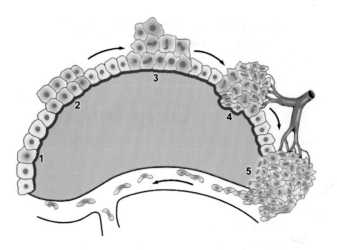

Fig. 5.1. Development of cancer

1. Mutation occurs in a single cell. Normally, this would be of no consequence as it would be readily removed by the host's defenses. In cancer patients, this does not happen and the cell survives; **2.** The mutated cell multiplies and produces a growth which at this stage is benign. This process is known as hyperplasia; **3.** Further multiplication occurs but abnormal cells have started to appear and this process is known as dysplasia; **4.** The tumor has now become large and many abnormal and rapidly multiplying cells have appeared. It has not yet broken any boundaries. At this stage it is called *in situ* cancer; and **5.** The tumor is now obviously malignant and has broken through the basement membrane and invading the neighboring blood vessels. The malignant cells are being carried via the blood to sites elsewhere in the body, producing metastases or secondaries. The tumor has developed its own blood supply which sustains its growth. This process is known as angiogenesis. The growth of cancer is also accelerated by the genetic instability of cancer cells. Successive clones continue to accumulate as cancer expands.

There are more than 100 forms of cancer but they could be broadly divided according to the tissues from which they emanate. Cancers arising from the epithelial tissues are called carcinoma, from the connective tissues, sarcoma, from the blood forming tissues, leukemia, from the lymphoid tissues, lymphoma, and from glanduler tissues adenocarcinoma. In addition, there are cancers which do not fall into these categories, i.e. the brain, nerve and skin cancers.

Genetic information carried by the DNA of the cell is not absolutely stable. Occasionally this changes, and the cell mutates. Cancers arise as a result of mutation, which may be acquired or inherited. Inherited or germ line mutations confer an increased risk of getting cancer; examples of this are the breast, ovarian, colon and prostate cancers. In addition, there are some precancerous hereditary diseases and one of these is polyposis coli in which numerous polypi appear in the colon (large intestine) which almost always change to adenocarcinoma.

In most cases, mutations are acquired, occurring at the tumor site, and there are three main causes for this. These are radiation, chemical exposure and infection. It has been known for many years that radiation can cause cancer. The use of X-rays and radioactive material for medical, industrial and research purposes increase the risk of cancer. Radiologists, for example, have a 7–10 fold greater chance of getting cancer than the general public. Survivors of the atomic bombs dropped on Nagasaki and Hiroshima have a much higher rate of cancer, as compared with residents of other Japanese cities. Tissues which are particularly susceptible to radiation are the thyroid gland, breast and bone marrow.

The sun's ultraviolet light is also a major cause of skin cancer in the white population, whose skin is not protected by melanin (skin pigment). The highest rate of skin cancer is seen in Queensland, Australia, because of the high intensity and duration of sunlight. UV light is not only mutagenic; it also suppresses the immune system, thus impairing its ability to destroy cancer cells.

Chemical carcinogens are probably the most important cause of cancer, on a worldwide basis. Numerous chemicals, on inhalation, ingestion and contact, can cause cancer. Cancers are not contagious, but a number of infectious agents, particularly some viruses, are capable of inducing cancer by integration into the host cell genome and activating the proto-oncogenes.

Immunosuppression, produced by certain drugs used to prevent rejection of organ transplants and human immunodeficiency virus (HIV), are associated with increased incidence of several cancers;

in particular, lymphoma and Kaposi's sarcoma, mostly seen in Africa.

Age and hormones also have a major impact on the development of cancer. Throughout life, cell mutants are produced, but they get destroyed by the host defenses. With age, these defenses get impaired and mutants are better able to survive. In the female, the rate of growth of the cancer of the uterus and breast appears to co-relate with estrogen secretion. Similarly, in the male, prostate cancer is depressed by estrogen and stimulated by testosterone.

More than 200 genes from different human cancers have been recognized which may provide a basis for the development of new means of diagnosis, treatment and epidemiology. In addition genetic markers in tumor cells have also been identified that are associated with metastases. Patients with positive markers may require more aggressive therapy.

Confirmatory diagnosis in cancer is based on histological examination of biopsy tissue. Biopsy is done by an operative procedure in which a bit of suspicious tissue is removed surgically. Alternately, aspiration of tissues can be done through a needle known as fine needle aspiration cytology (FNAC). This is minimally invasive and gives quick diagnosis.

Body scanning in the form of magnetic-resonance imaging (MRI) and computerized tomography (CT) provide excellent diagnostic information which was not available previously. Amongst the latest imaging techniques is positron emission tomography (PET). Positrons are tiny particles emitted by radioactive substance administered to the patient along with glucose. As cancer cells metabolize glucose at a faster rate the radioactive signals are stronger wherever the cancer cells are present. This technique is especially useful in the detection of metastasis or secondary tumors.

In most cases of cancer, surgery is the preferred treatment. However, there are a few problems associated with surgery. An operative procedure, however cannot always ensure the total removal of all cancer cells. Surgery is also difficult if vital organs are involved and metastases have occurred.

Cancer cells are susceptible to radiation; therefore radiotherapy is widely used in the form of powerful X-rays or gamma rays. It was initially thought that radiation directly killed the malignant cells but we now know that the radiation-caused DNA damage is relatively minor, and that the cells kill themselves by a self-destructive process known as apoptosis (see Chapter 24).

Chemotherapy, with cytotoxic drugs is very effective in the treatment of some cancers, especially when given as a cocktail. Leukemias, lymphomas and testicular cancers are now successfully treated by this method. Unfortunately, the chemotherapeutic agents often produce side effects because they also kill some healthy cells, such as the rapidly growing bone marrow cells, leading to anemia and immunodeficiency.

However, there are now promising new drugs targeting enzymes that signal proliferation of cancer cells, thereby sparing the normal cells. The most well known of these drugs is Gleevec (Imatinib Mesylate), used for the treatment of gastrointestinal stromal tumors (GIST) and chronic myeloid leukaemia (CML). A number of other similar drugs are in the pipeline as well.

Two promising developments have occurred in cancer immunology. The first is an antibody-based therapy and the second is immunization, using cancer antigens. In the first procedure, antitumor agents are attached to monoclonal antibodies[b] and are then given to the patient. Unlike standard chemotherapy, the toxic chemical is delivered directly to the cancer cells, leaving the healthy cells unharmed. In the second procedure, the patient is given a vaccine to stimulate his immune system so as to produce antibodies against the cancer cells. This is known as therapeutic vaccination and several trials are going on at present.

However, the best safeguard against cancer is prevention as most cancers are avoidable if proper precautions are taken. Recent studies indicate that Aspirin and related anti-inflammatory drugs, may have a preventive role in certain cancers although it is not known why this happens. Reports also indicate that frequent

---

[b]Antibody which reacts only with a single component of the immunogen.

drinking of green tea is inversely associated with the risk of developing several types of cancers. This is due to a chemical in green tea, epigalloctechin-3-gallate (EGCG), which counters another chemical vascular endothelial growth factor (VEGF), which is crucial for the survival of tumor cells. Similarly, chrysanthemum flower extract appears to inhibit proliferation of various cancer cell lines and may be of potential benefit in prevention.

A List of Known Chemical Carcinogens

Substance	The Cancer it is Suspected to Cause	Its Source
Aflatoxin	Liver cancer	A toxin produced by fungus growing on nuts.
Aniline dyes	Bladder cancer	For dying textiles, wood, etc.
Arsenic	Skin cancer	Pesticides and wood preservatives.
Asbestos	Lung cancer	Used as roofing and insulating material.
Benzene	Lung cancer	Paint.
Betel nut	Mouth	Chewed in some countries, particularly South Asia.
Bracken (fern)	Stomach (if ingested)	A member of the *Pteridium* family of ferns. Widely distributed.
Diesel fumes	Lung cancer	In poorly ventilated spaces.
Dioxins	Breast and liver cancer	Burning of plastics. Gets stored in animal fat and appears in dairy products.
Heterocyclic amines	Stomach cancer	By over-cooking and charring of food.
Tar	Skin cancer	Coal tar.
Tobacco smoke	Lung cancer	In cigarette, cigar and pipe smoke.

Infections Associated with Human Cancers

Viruses	Associated Tumor	Areas of High Incidence
DNA Viruses		
Hepatitis B virus (HBV)	Liver cancer	Southeast Asia, tropical Africa
Human Papilloma virus (HPV)	Carcinoma of cervix	Worldwide
Epstein–Barr virus (EBV)	Burkitt's lymphoma, nasopharyngeal cancer	West Africa South China
Human herpes virus-8 (HHV-8)	Kaposi's sarcoma	Africa
RNA Viruses		
Human T-Cell leukemia virus type 1 (HTLV-1)	Adult T cell Leukemia/lymphoma	Japan
Human immunodeficiency virus (HIV-1)	Kaposi's sarcoma	Central Africa
Hepatitis C virus (HCV)	Liver cancer	Japan, Italy, Spain and South Asia
Bacteria		
Helicobacter pylori	Causes atrophic gastritis, a possible precancerous condition	Worldwide
Parasites		
Opisthorcis viverrini	Biliary cancer	Thailand
Schistosoma hematobium	Bladder cancer	Egypt
S. japonicum	Colon cancer	China

Only a small number of individuals infected by these organism get cancer. This means that the infection acts in conjunction with *genetic susceptibility* to produce the disease.

Differentiation between Benign and Malignant Tumors

Characteristics	Benign	Malignant
Growth	Slow and expansive, often encapsulated	Rapid and invasive, non-encapsulated.
Recurrence after removal	Absent or rare	Frequent, if not completely removed.
Histology	Resemblance to normal cells Mitoses infrequent	Abnormal cells. Mitosis frequent.
Metastases	Absent	Present (If treatment is delayed).
Systemic or constitutional effects	Rare (unless endocrine glands are involved)	Common (loss of weight, fever, etc.).

Preventive Measures*

1. Do not smoke. This is the most important preventable cause of cancer.
2. Be moderate in the consumption of alcoholic drinks.
3. Avoid excessive exposure to sun.
4. Avoid exposure to carcinogens and follow safety instructions when handling toxic substances.
5. Frequently eat fresh fruit, vegetables and cereals with high fiber content.
6. Avoid becoming overweight and limit the intake of fatty foods.
7. Have a cervical smear regularly.
8. See a doctor if you notice a lump, abnormal bleeding, changes in a mole, changes in bowel habits, have persistent cough or unexplained weight loss.
9. Check your breasts regularly and if possible have a mammogram at regular intervals after the age of 50.

Recommendations based on the European code against cancer.

Percival Pott

Sir Percival Pott, an English physician was first to put forward the concept of carcinogen in 1775. At that time young boys were used as chimney sweeps as they were small enough to climb inside and clean the soot. Pott noticed that the boys had an unusually high rate of scrotal cancer. He suggested that the causative agent is likely to be the soot and recommended frequent washing and changing of clothing so as to reduce the exposure to soot. His observation also demonstrated that cancer may take many years to develop after exposure to the causative agent. We now know that soot has tar which is a potent carcinogen.

❋❋❋❋❋

"One thing I have learned in a long life,
that all our science, measured against reality,
is primitive and childlike — and yet it is
the most precious thing we have".

Albert Einstein

6

Extending the Range of Vision

Philosophers have long wondered if it is possible to conceive of any object (in the physical sense) beyond our perception? This is a debatable question, but it is correct to say that we cannot go beyond the images stored in our memory bank. Of course, new objects get added to this bank, but we have to perceive them first.

Let us take, for example, the black hole. Nobody has seen a black hole, but we have seen black and holes, so in our mind we can construct an imaginary black hole. In reality, the black hole may be neither black nor a hole.

We can conceive of God, but this too is based on human qualities with which we are familiar, and which lie within our perception. When Michelangelo made his magnificent painting on the roof of the Sistine Chapel, he had no choice but to make God in man's image!

Similarly, before the invention of the light microscope, humans had no idea of microbes, or at least of their physical appearance. Viruses too were beyond our conception, until the invention of the electron microscope (Fig. 6.1). In fact, Beijerinck (1898), who first used the term virus, thought they were liquid!

A Dutchman, Anton van Leeuwenhoek (1632–1723), is credited with the invention of the light microscope, but microscopes were being used prior to his birth. They were, however, only enlarging objects about twenty or thirty times their natural size. Leeuwenhoek very skillfully ground the lenses and was able to

Resolving power of microscopes

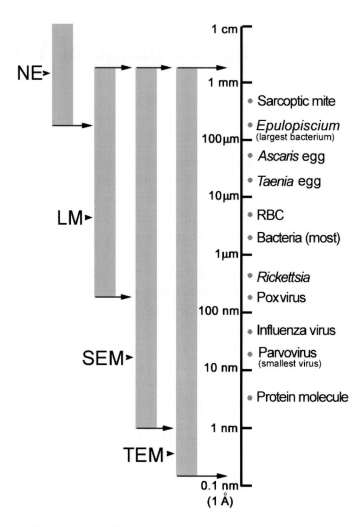

Fig. 6.1. Microorganisms show great variations in size. Some of these are shown along the scale for comparative purpose. Parvovirus (22 nm) is amongst the smallest of viruses and poxvirus amongst the largest (250 nm).

NE = Naked eye　　　　　　　　　　　LM = Light microscope
SEM = Scanning electron-microscope　　TEM = Transmission electron-microscope

build microscopes which could enlarge objects up to 200 times their normal size. With this microscope, Leeuwenhoek started examining a variety of tissues. He first saw protozoa in water and sent a description of a ciliate (*Vorticella* sp), to the Royal Society of London. He was also the first to see bacteria, which he collected from dental plaque, and called them "animalcules". In one of his letters he wrote: "...my work which I've done for a long time, was not pursued in order to gain praise I now enjoy, but chiefly from craving after knowledge". The quality of a true scientist!

With the passage of time, microscope design kept on improving. In 1876, Abbe analyzed the effects of diffraction on image formation and re-designed the microscope. In 1886, Zeiss made high quality lenses, and using Abbe's design, made microscopes almost reaching the theoretical limits of visible light.

The limitation on the resolution of an object, under the light microscope, is because of the wavelength of visible light, which is between 0.4 μm (micrometer) to 0.7 μm. Any object smaller than 0.4 μm becomes difficult to resolve. The limit of resolution is defined as a minimum distance between two closely positioned objects, at which the two objects can be observed as distinct entities, and this in the best light microscope is no more than 0.2 μm.

Phase-Contrast microscopy utilizes differences in refractive indexes between various cell structures. This enables clearer visualization because of the increased contrast. For example details of the nuclear structure could be seen.

Interference microscopy separates light into two beams, one of which is passed through the specimen and then is combined with the other. The phase shift between the two beams, produces an interference pattern, resulting in structures to appear in different colors, increasing the contrast and giving the specimen a three-dimensional appearance.

Polarizing microscopy uses polarized light and structures which are able to alter the behavior of the polarized light, such as crystals, to show birefringence. This is useful in recognizing crystalline material in tissues.

Darkfield microscopy uses light aimed at the sample from an angle. Only those light rays deflected by the object are visualized against a dark background. Small objects such as spirochaetes (very thin bacteria), which are not visible in the ordinary light become visible.

Fluorescent microscopy involves the use of a light source which emits light in or near the ultraviolet (UV) range. The fluorescent microscope is fitted with two sets of filters; an excitor filter in front of the light source and a barrier filter in the eye piece. The excitor filter permits only selected wavelengths from the light source to pass through on the way to the specimen. On striking the specimen containing a fluorescent dye, an altered wavelength is emitted depending on the fluorescent dye used. For example, fluorescein will emit green light and rhodamine red light. The barrier filter cuts off unwanted fluorescent signals and protects the eye from UV light. (Front cover page shows specimen stained with a fluorescent dye)

A fluorescent dye can be conjugated with an antibody and used as a probe for the detection of specific antigen. This technique is called *imunofluorescence* and is widely used for diagnosing microorganisms and antigen in tissues.

Confocal microscopy uses laser to focus on a single point at a specific depth in a specimen. The focus can be moved from point to point, resulting in "optical sectioning". From a series of such optical sections, the attached computer is able to build three-dimensional images of the object being examined.

Electron microscopy gives a much higher resolution than the light microscope because at an acceleration voltage of 50 kv, a wavelength of 5×10^5 times shorter than that of the visible light is produced. The illumination source is the electron gun, which produces a stream of electrons. The electrons pass unhindered in a vacuum. It is, therefore, essential that the entire column housing the various parts of the microscope be maintained in a vacuum during operation. Most of the electrons pass through the specimen that is being examined. However, some electrons get scattered by heavy atoms in the specimen and deviate from the beam. This produces an image which can be seen on a fluorescent screen. This type of electron microscope is known as the transmission electron microscope or TEM, in contrast to the scanning electron microscope or SEM.

In SEM, an ultra fine beam scans the surface of the specimen, which is coated with a heavy metal. The quantity of electrons scattered is measured and used to control a second beam which moves in synchrony with the first, and forms an image on a fluorescent screen. SEM provides a three-dimensional image and has a very wide range of magnification. It is, therefore, a very useful instrument to study the surface topography of any specimen (Fig. 6.2).

Good specimen preparation technique is very important for achieving best results from electron microscopy. The two most commonly used preparation techniques for TEM are negative staining and thin sectioning of resin embedded specimen. Negative staining is used principally for examining viruses and bacteria. In this technique, the specimen is first dropped on metal grids and then covered with stain containing heavy metal salts.

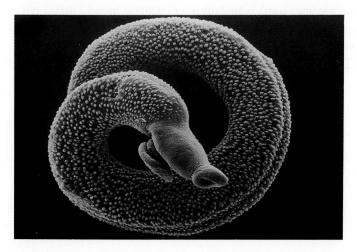

Fig. 6.2. SEM of *Schistosoma mansoni*, a trematode worm which lives inside the blood vessels. Note the three-dimensional image and the elucidation of the surface pattern. ×5000.

The stain gives a dark background against which the specimen stands out, revealing its surface structure.

In thin sectioning, an ultramicrotone is used to cut through the cell or other tissues, revealing their *internal* structure. This has been the major contribution of electron microscopy to the field of biology and has enabled us to understand how the cell functions and the involvement of organelles in those functions.

Physical Measurements

10 Angstroms (Å) = 1 nanometer (nm)
1000 nanometers = 1 micrometer (μm)
1000 micrometers = 1 millimeter (mm)
10 millimeter = 1 centimeter (cm)
100 centimeter = 1 meter (m)

Measurements Equivalents

1 centimeter (cm) = 10^{-2} meter (m) = 1/100 m = 0.4 inch
1 millimeter (mm) = 10^{-3} m = 1/1000 m = 1/10 cm
1 micrometer (μm) = 10^{-6} m = 1/1 000 000 m = 1/10 000 cm
1 nanometer (nm) = 10^{-9} m = 1/1 000 000 000 m = 1/10 000 000 cm
1 meter = 10^2 cm = 10^3 mm = 10^6 μm = 10^9 nm

✳✳✳✳✳

"The gift of the great microcopist is the ability to think with the eyes and see with the brain".

Daniel Mazia
Cell Biologist

7

On Life's Fringes

In 1892, D. Ivanowsky of Russia, showed that the tobacco mosaic disease, a crippling disease of the tobacco plant, was caused by an agent which passes through a bacterial filter, which meant that it was smaller than the bacteria. In 1898, two German scientists, Friedrich Loeffler and Paul Frosh presented evidence that the causal agent of the foot-and-mouth disease of animals, was also filterable. They were certain that this substance was not a toxin produced by bacteria, as it produced sickness in animals even on injection of a small dose, and the disease was transmissable from animal to animal, which cannot happen with a toxin. The above discoveries established that plants and animals get infected by filterable agents, which were given the name of virus (latin = poison), by M.W. Beijerinck (1898), professor of microbiology in Delft, Netherlands. Another amazing discovery later followed and an English man, F.W. Twort (1915), and a French scientist F.d' Herelle (1917), discovered that bacteria too have viruses infecting them! They named the bacterial virus, bacteriophage (Gk, Phagein = to eat), i.e. organisms which eat bacteria (Fig. 7.1).

Viruses are generally not regarded as "living" because they are unable to reproduce on their own. However, viruses are extremely efficient parasites and recruit the resources of the host cell to produce their progeny (Fig. 7.2). This remarkable achievement is unique in the field of biology.

Bacteriophage

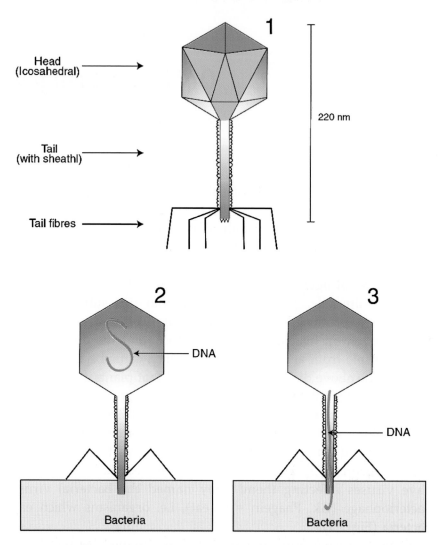

Fig. 7.1. Showing the method of entry of viral DNA into bacteria. The viral DNA is injected through the phage tail into the bacterium, with the viral protein shell remaining outside. The phage DNA instructs the bacterial machinery to synthesize phage components. The phages so produced are released when the bacterium undergoes lysis. Phages are highly host specific, they attack only certain strains or sub-groups within a species. The phages can, therefore, be used for typing strains. This is known as phage typing.

VIRAL LIFE CYCLE

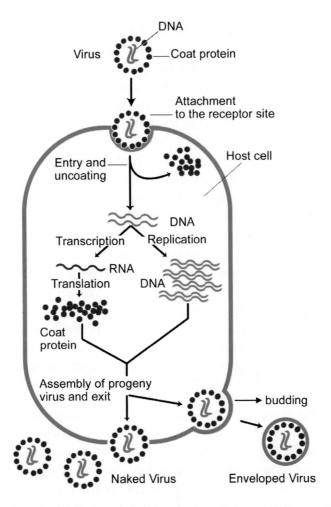

Fig. 7.2. Viruses first attach to specific receptor sites on the host cell surface and then gain entry into the cell. They may replicate in the nucleus or cytoplasm and exit by lysis of the cell or by budding.

Replication = the doubling of a nucleic acid molecule to form two identical products.

Transcription = the rewriting of the genetic message of DNA into RNA.

Translation = the transfer of genetic message from RNA to the corresponding amino-acid sequence of a protein.

As there are numerous viruses, they need to be properly classified. The following criteria are used for this purpose:

1. **The nature of nucleic acid in the virus genome:** This could be either RNA or DNA, but never both.
2. **The symmetry of the capsid:** (the outer protein coat which protects the genome). The capsomeres (subunits of the capsid), can show either cubic symmetry, in which case capsomeres form an icosahedron, with 20 identical triangular faces, or they can show helical symmetry. In the latter case the capsomeres are spirally arranged, forming a hollow cylinder.
3. **The presence or absence of an envelope around the capsids:** The envelope is made up of proteins, lipids and carbohydrates and believed to be derived from the host cell membrane. The viruses with the envelope are known as "enveloped viruses" and those without as "naked viruses".
4. **The size of the virus:** Their size varies from 20 nm (nanometer) to 300 nm (1 nm = 10^{-9} m). The large viruses are almost the size of a small bacterium.
5. **Nucleic acid sequence:** For comparison, gene sequences are available in databases for the prototype viruses of nearly all the taxa or major groups. Sequencing is a very accurate method of identifying and establishing their relationships.

The virus morphology is studied with the electron-microscope, often using the negative staining technique. In this technique a suspension of virus particles is put on a grid and washed in 1% aqueous potassium phosphotungstate, which is electron dense. After drying, the phosphotungstate covers the grid, except where the viruses are located. Because the virus particles allow the electron to pass through, in contrast to the background, a clear negative image of the virus is obtained (Figs. 7.3 and 7.4).

As the viruses need live cells to grow, they are maintained in tissue culture, which are cells grown *in vitro* (literally "in glass"). Numerous cell lines are now available to suit the requirements of a particular virus. When a virus grows in tissue culture, it produces cytological change, which is known as cytopathic effect

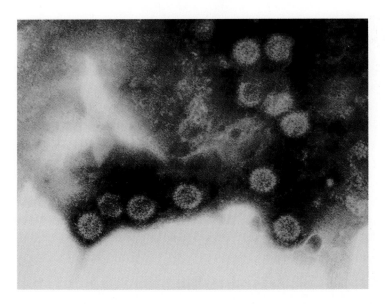

Fig. 7.3. Negative staining of Rota virus from feces of a diarrhea case. The tiny spike like structures surround each virus particle. Negative staining. ×265 000. TEM. (Courtesy of Ms. Josephine Howe, Department of Microbiology, NUS)

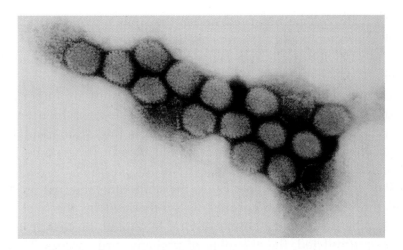

Fig. 7.4. Negative staining of adenovirus showing symmetrically arranged capsomeres, visible as small dots. This is an example of a naked virus. Negative staining. ×150 000. TEM. (Courtesy of Dr. Mary Ng, Department of Microbiology, NUS)

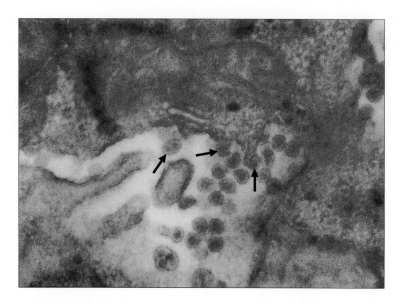

Fig. 7.5. Section of tissue culture cells infected with West-Nile virus. These are enveloped and can be seen budding out of the host cell. ×130000. TEM. (Courtesy of Dr. Mary Ng, Department of Microbiology, NUS)

(CPE). CPE can be observed under the light microscope and is often characteristic of a particular virus.

The viral life cycle in a cell generally lasts between 6 to 36 hours. The parental virus particle first adsorbs to specific receptors located on the host cell surface. The receptor sites are important, as different viruses attack different cells. The virus then enters the cell by a process similar to phagocytosis (ingestion displayed by macrophages and white cells of the blood). Its genome gets uncoated, and it replicates, either in the cytoplasm or the nucleus, depending on the virus species. At the same time, the coat protein that the virus genome encodes, is synthesized by the host cell RNA. When sufficient amounts of viral DNA and the coat proteins have accumulated, the assembly of progeny virus occurs. This is followed by the exit of the virus from the cell (Figs. 7.2 and 7.5).

Viruses can spread by multiple routes; these include inhalation, ingestion, direct contact, injection, sexual intercourse and arthropod and animal bites. A few viruses also pass from the mother to the child by the congenital route, i.e. via the placenta. This is known as vertical transmission.

Viruses transmitted by the aerosol route are very common. They spread rapidly by coughing and sneezing, affecting large population groups in a few days. Most respiratory viruses produce mild diseases but some of them such as influenza[a] can have a devastating effect on the human population. In 1918–1919, the influenza pandemic caused at least 20 million deaths. Influenza epidemics usually occur every few years because of the change in the virus's outer protein coat, to which the population has little or no immunity. When the antigenic change is relatively minor, it is called "antigenic drift"; when it is major, it is called "antigenic shift". Shifts are more dangerous, as the body is totally unprepared to handle what is virtually a new virus. Shifts occurred in 1918, 1957 and 1968.

Influenza viruses also infect animals including birds, pigs and horses. There are about 15 avian or bird strains which remain largely confined to birds. However, a recent avian strain, (H5N1),[b] surpassed the species barrier and infected humans in S.E. Asia and the Far East. This led to severe illnesses in humans with some deaths. Thousands of chickens had to be culled to control the spread. Exchange of genetic material between bird and human influenza virus can also occur. This process is called reassortment, resulting in a new strain capable of infecting humans and causing epidemics.

Amongst the respiratory infections, the most alarming in recent years has been the Severe Acute Respiratory Syndrome (SARS). The infection first emerged in South China in November, 2002. The major clinical features of the disease were persistent fever,

---

[a]The word influenza comes from the ancient idea that the disease is caused by "divine infuence".
[b]The letters H and N represent surface proteins on the virus. H is haemagglutinin and N is neuraminidase.

body pain, chills, dry cough, headache and breathlessness. X-rays showed pneumonia in most cases. The Chinese physicians initially labeled it as "atypical pneumonia" and preventive measures were not taken, resulting in the infection rapidly spreading to other parts of China and Hong Kong. It further spreads to Vietnam, Singapore, Taiwan and Canada, via passengers acting as carriers or agents of this infectious disease.

It was soon realized that SARS is a highly contagious disease and the infectious agent is discharged from the respiratory system in droplets. Droplets in a sneeze can travel as fast as 150 ft in a second and as far as 12 ft. Although it had an overall mortality rate of 9.6%, the disease was eventually brought under control using strict quarantine, temperature monitoring of passengers, contact tracing and the use of protective measures by health givers (e.g. hand washing, gloves, gown, eye protection and N95 face mask).

Finally, the causative agent was isolated from patient tissues and it turned out to be a coronavirus. The coronavirus is so-called because it has petal shaped projections giving it the appearance of solar corona (Fig. 7.6). SARS CoV grows luxuriently in Vero cells[c] and this prolific growth pattern is unlike other human coronaviruses (Figs. 7.7, 7.8 and 7.9). This may account for the rapidity of symptoms that develop in SARS. Several laboratories have now sequenced the virus genome and the SARS CoV is not related to any other coronavirus known previously. It became obvious that the infection which appeared for the first time in humans, had an animal origin.

Yi Guan and colleagues from China unravelled the mystery by isolating SARS CoV from Himalayan civet cats (*Paguna larvata*), sold in the wet markets in Shenzen, China, and eaten as a delicacy. The researchers also found serological evidence for SARS CoV in people working in the wet markets and some restaurant workers.

---

[c]Green monkey kidney cell line.

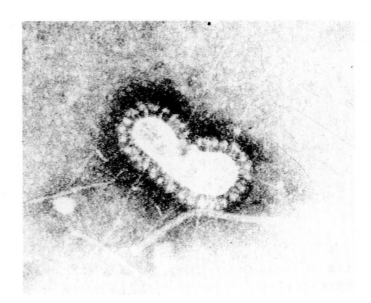

Fig. 7.6. Corona type virus to which SARS virus belongs. Negative staining. ×200 000. TEM. (Courtesy of Dr. John Marshall)

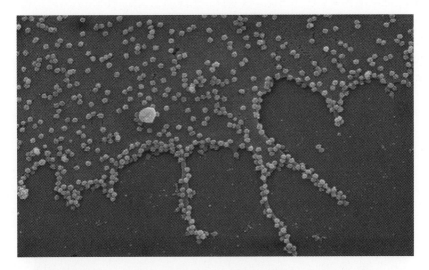

Fig. 7.7. SARS coronavirus growing in Vero cells. Numerous mature virus particles are seen on the cell surface and along the margins of the infected cell. ×20 000. SEM. (Courtesy of Ms. Micky Leong, EM unit, NUS and Dr. Mary Ng, Department of microbiology, NUS)

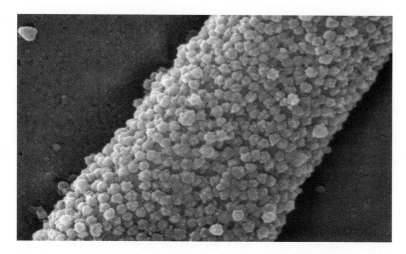

Fig. 7.8. SARS coronavirus growing in Vero cells. High magnification showing extensive numbers of virus particles on the cell filopodia at 15 hours post infection. This prolific growth is unlike other human coronaviruses. ×40000. SEM. (Courtesy of Ms. Micky Leong, EM unit, NUS and Dr. Mary Ng, Department of microbiology, NUS)

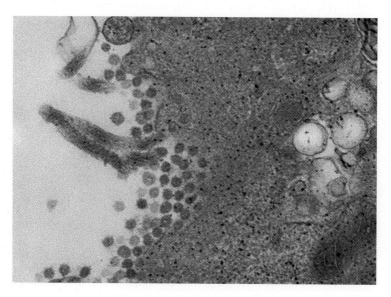

Fig. 7.9. SARS coronavirus growing in Vero cells. Mature virus particles are seen in large numbers at the infected cell margin. ×40000. TEM. (Courtesy of Ms. Tan Suat Hoon, EM unit, NUS and Dr. Mary Ng, Department of microbiology, NUS)

This showed that SARS is a zoonotic disease (disease transmissable from animals to humans). The future control will depend on breaking the animal to human transmission and monitoring the wet markets to prevent the sale of civet cats or any other animal harboring the virus.

Smallpox, the most dreaded of all the ancient diseases, is caused by a large DNA virus belonging to the poxvirus group (Figs. 7.10 and 7.11). It is contracted by the inhalation of the virus, and by direct contact with the scabs shed by the patients. The eradication of the disease was achieved because the virus had no animal reservoir, and did not produce an undetectable or asymptomatic carrier state. In addition, there was a highly effective vaccine available, which gave solid life-long immunity. Most of the remaining poxviruses produce a mild disease. This includes molluscum contagiosum, a disease which involves the epidermal layer of the skin, producing small umblicated benign tumors.

The viruses transmitted by arthropods are also known as arboviruses, being the short form of *ar*thropod *bo*rne viruses. The arthropods involved are mosquitoes, ticks and sandflies. There are more than 400 such viruses, of which about 60 infect humans, the rest involving domestic and wild animals. Arboviruses, unlike the smallpox virus, generally have a large animal reservoir, which makes their control difficult. The animal reservoir includes birds, pigs and horses. The viruses do not necessarily produce disease in these animals, but they act as the "amplification host", i.e. viruses multiply in their tissues enabling easy transmission to arthropods.

Arboviruses are the main agents causing "Hemorrhagic fevers", in which the patients develop hemorrhages all over the body and bleeding occurs via the orifices. Arboviruses also cause encephalitis or inflammation of the brain, and in some cases, inflict severe damage to the liver, resulting in jaundice or yellow colored skin, hence the name of yellow fever given to this disease. Yellow fever was originally confined to Africa but was brought to the Americas during the slave trade. This is an excellent example of how human activity can result in the spread of infectious diseases.

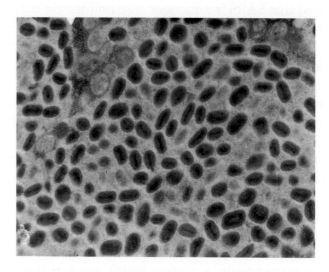

Fig. 7.10. *Molluscum contagiosum* virus belongs to the family Poxviridae. These are the largest amongst the viruses which are brick-shaped. Smallpox, which has been erradicated, belongs to this group. ×20 000. TEM.

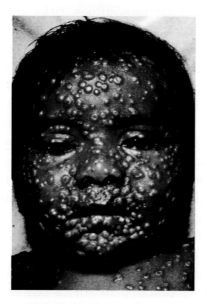

Fig. 7.11. Smallpox in a 9-month old boy (Courtesy of WHO). This is the only disease which has been totally eradicated. The last case of smallpox was recorded on October 26, 1977 in Somalia.

The sexual route of transmission is utilized by a number of viruses and the most important of these is the human immuno-deficiency virus (HIV), which causes the acquired immuno-deficiency syndrome (AIDS). AIDS was first described in 1981 in homosexual men in N. America. The discovery of AIDS in Africa in 1983, among those who were neither homosexual nor IV drug users, led to the realization that HIV is transmissable by heterosexual intercourse as well. The disease probably existed in Africa long before 1983, as was shown by the presence of HIV antibodies in a serum sample from Congo that was stored in 1959. HIV belongs to a large group of viruses known as retroviruses. They are so-called because of the unique feature of being able to produce complementary DNA from a RNA template. They do this by the possession of an enzyme called reverse transcriptase.

HIVs are genetically similar to viruses found in the African primates, called SIVs, the simian immunodeficiency viruses. Therefore, it is postulated that HIVs are mutants of SIVs but it is not clear as to how SIVs got transmitted from primates to humans.

HIVs have selective tropism (attraction), for CD4 molecules found on the surface of T helper cells, monocytes and macrophages. As these cells help in maintaining the immunological defenses of the host, their destruction exposes the AIDS patients to many opportunistic infections. Death in AIDS usually occurs because of secondary infection caused by fungi, bacteria, and parasites.

The viruses that spread by ingestion can give rise to diarrhea and other intestinal problems e.g. Rotavirus (Fig. 7.3). In addition, some viruses can cause inflammation of the liver or hepatitis. The poliovirus also spreads by the oral route, but does not cause intestinal disease. It causes destruction of the spinal nerves, resulting in paralysis. There are mainly 5 hepatitis (A, B, C, D, E) producing viruses which are very common in the developing countries. Some are acquired by the oral route (HAV, HEV), while the remaining three (HBV, HCV, HDV) are acquired through injection of blood, blood products and body fluids. HBV and HCV can produce chronic infection and permanent liver damage, including cancer.

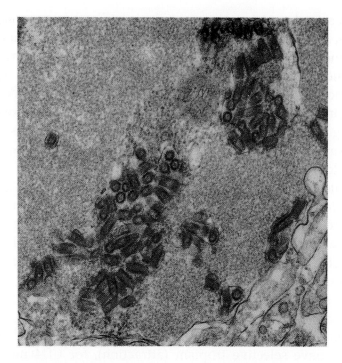

Fig. 7.12. Rabies virus belongs to the family Rhabdoviridae. When seen longitudinally they are bullet shaped. In cross-section they appear circular. This is an infected mouse brain and the virus has caused degeneration of the brain cells in its vicinity. ×80 000. TEM.

Amongst the viruses transmitted by animal bites, rabies is by far the most important. The disease was known in India and the name rabies comes from the Sanskrit word *rabhas*, meaning violence. The rabies virus is a bullet-shaped RNA virus (Fig. 7.12) which selectively invades the nervous system. It is introduced into the tissues via the saliva of the rabid animals, mostly belonging to the dog family. In some parts of the world, vampire bats are also involved.

Following the deposition of the rabies virus in the bite wound, it enters the peripheral nerve fibers and ascends towards the spinal cord, where it undergoes the first replication cycle. The virus then goes into two directions, some particles returning to the sensory fibers at the site of the bite while other particles go up the spinal cord to the brain. The return to the site of the bite accounts for the itching that commonly occurs, 3 or 4 days after the infection. The virus invading the brain concentrates in the limbic system (where emotional centers are located). After the virus has replicated in the brain, it spreads centrifugally to various organs, including the sub-maxillary salivary glands. From these glands it is discharged in the saliva. The involvement of the limbic system probably accounts for the development of aggressive behavior in otherwise docile dogs and produces what is known as "furious rabies". This behavior change in the carnivores is the key to the survival of the virus both in the wild and in the human habitation.

The treatment of viral diseases is not easy, because the viruses get intimately involved with the host cell. To selectively attack the virus, without damaging the host cell, is difficult, and this is the reason why not many anti-viral agents are available as compared to other microbes and parasites. However, prevention against many viral diseases by immunization has been very successful, and, like smallpox, at least a few other viral diseases are likely to be eradicated in the near future.

Hela cells

This is the most famous cell line used for culturing viruses. It was obtained from the cervical cancer of a patient, Henrietta Lacks, more than fifty years ago. Henrietta had come to Johns Hopkins Hospital in Baltimore for the treatment of cancer. The gynaecologist, during the process of inserting radium needles in her cancerous tissue, took a small bit of cancer and sent it to Dr. George Gey, head of the tissue culture research. Henrietta died a few months later but her cells grew in the tissue culture and are still living today. They are used by numerous labs all over the world for growing viruses. Henrietta is dead but her cells have become immortal!

✳✳✳✳✳

"Observe the virus: think how small
Its arsenal, and yet how loud its call;
It took my cell, now takes your cell,
And when it leaves will take our genes as well".

Michael Newman
The Sciences, 1982

8

A Friend and a Foe

The Earth's fossil records show that bacteria appeared more than 3.5 billion years ago and enabled the development of other life forms by providing oxygen to the atmosphere. If all the bacteria were to die today, life will cease to exist on earth. If one judges the evolutionary success by the yardstick of survival, bacteria are indeed highly evolved, remembering that in evolution there is no inexorable trend towards higher forms of life. They have survived extremes of temperature and climate for billions of years and continue to exist as the most prolific forms on Earth. Their success is at least partly attributable to "lateral gene transfer" — obtaining genes from each other. They can acquire gene sequences through conjugation (DNA transfer as a result of bacterial mating) (Fig. 8.1), transformation (when bacteria pick up DNA which is no longer part of another organism) and transduction (in which bacterophage transfer DNA fragments).

Bacteria are to us, both a friend and a foe. Let us first look at the help and assistance they provide. Inside the intestinal tract they produce enzymes and vitamins. The enzymes are particularly important for the herbivores, as without them, the plant cells could not be broken down. If one kills all the resident microbes in the termite intestines, the termites die because the wood splinters remain undigested. The bacteria are highly efficient scavengers. They break down all the organic matter that falls on the ground, so that dead animals and plants release their valuable chemicals for use by the other organisms. In fact, bacteria have been identified which can degrade even heavy metals such as iron.

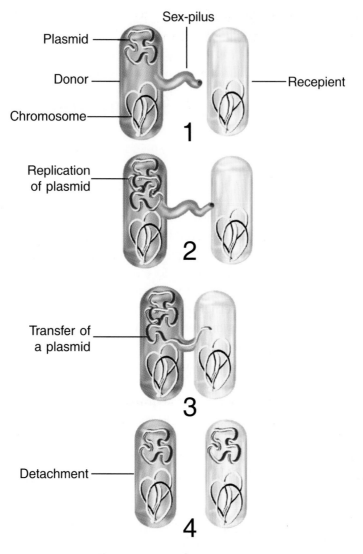

Fig. 8.1. Bacterial Conjugation

1. The donor bacterium produces a sex-pilus to connect it to the recipient bacterium; **2.** Replication of the plasmid (extra-chromosomal DNA) occurs; **3.** A plasmid is transferred via the bridge made by the sex-pilus to the recipient; and **4.** The sex-pilus is withdrawn and the bacteria detach (Adapted from various sources).

Antibiotic resistance, carried on the plasmid, is transferred from one species to another by this method.

The nitrogen cycle, essential for life on earth, is maintained by bacteria. Plants are unable to use gaseous nitrogen from the atmosphere. The principal way in which nitrogen becomes available to them is through nitrogen-fixing bacteria, such as *Rhizobium*. These bacteria are able to convert gaseous nitrogen into nitrates and release them into the soil which the plants then utilize. Some plants, such as legumes, have evolved structures which can house these useful bacteria in their tissues. In return, plants supply carbohydrates to the bacteria. This is an excellent example of two different life forms coexisting in a symbiotic relationship.

As bacteria have a short life span, they provide us with an excellent tool to study the nature of life and the laws that govern it. With the help of recombinant DNA technology (genetic engineering), bacteria are now used for producing important

---

Some Bacterial Products of Recombinant DNA Technology

	Substance	Bacterium	Use
1.	Human insulin	*Escherichia coli*	Diabetes
2.	Human growth hormone	*E. coli*	Treatment of growth defects
3.	Tumor necrosis factor	*E. coli*	Killing tumor cells
4.	Epidermal growth factor	*E. coli*	Treatment of burns
5.	Interleukin-2	*E. coli*	Treatment of cancer
6.	Cellulase	*E. coli*	Breaking cellulose
7.	Pro-Urokinase	*E. coli*	Thromboloysis (dissolving clots)
8.	Porcine growth factor	*E. coli*	Weight gain in pigs
9.	Bovine growth factor	*E. coli*	Weight gain in cattle
10.	Interferons	*E. coli*	Viral infections
11.	Steptokinase	*Streptococcus*	Thrombolysis

medicinal substances such as hormones. For a long time, insulin, used by diabetics, was extracted from bovine or porcine pancreas. Now it is produced by bacteria. The harnessing of bacteria in this respect is no less important than the domestication of wild animals, which took place 9000 years ago.

Bacteria are found on our skin, in the nose, mouth and vagina. In the intestinal tract, the bacterial count progressively increases from the stomach downwards, reaching its maximum in the colon (large intestine). The high acidity in the stomach (pH 3), prevents most bacteria from surviving, but a spiral bacterium, *Helicobacter pylori* lives in it, sometimes causing ulcers. In view of this *H. pylori* is of great medical importance.

There are so many bacteria in, and on our body, that about 10% of our body weight consists of them. It has been estimated that the number of *Escherichia coli* (a species living in the gut) in a single individual, would be greater than the total number of humans that have ever lived on earth. Similarly, each square centimeter of our skin harbors an average of 100000 organisms. This is, however, normal, and not suggestive of any disease.

Washing with soap (scrubbing), reduces their number, but they reproduce so quickly that the original number is usually restored within a few hours. The normal flora of bacteria on the skin and in the intestinal tract protect us from other dangerous organisms. The use of antibiotics can disturb this balance and make us more prone to infections.

Bacteria range in size from 0.1 μm (micrometer) to about 600 μm. The largest bacteria so far discovered is the *Epulopiscium fishelsoni*, found in the gut of the sturgeon fish, which lives in the Red Sea and off the coast of Australia. *E. fishelsoni* is so large that it can be seen by the naked eye, appearing as a small dot.

Bacteria come in three main shapes. Rod-shaped called bacilli, round-shaped called cocci and spiral-shaped called spirochetes. All bacteria divide by simple binary fission in which the mother cell separates into two equal daughter cells (Figs. 8.2 and 8.3).

Most bacteria are fairly stationary but some are actively motile and move about by means of long whip-like structures called

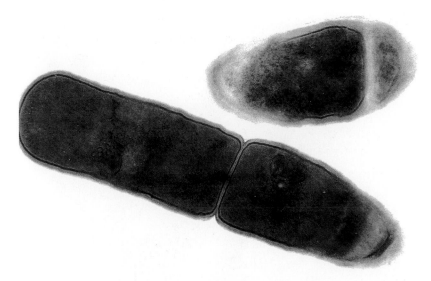

Fig. 8.2. Dividing bacteria. ×30000. TEM.

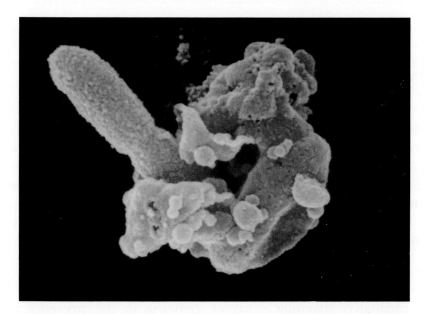

Fig. 8.3. A dividing bacterium being ingested by a macrophage. ×25000. SEM. (Courtesy of Dr. P. Gopalakrishnakone and Ms. Chan Yee Gek, Department of Anatomy, NUS)

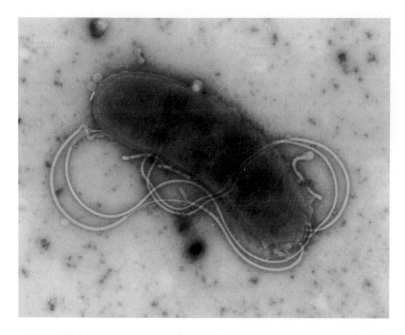

Fig. 8.4. Bacteria move by using whip-like organelles known as flagella. In this case negative staining has been used to demonstrate the flagella in *Helicobacter pylori*, a bacterium which lives in the stomach. ×52000. TEM. (Courtesy of Dr. Ho Bow and Ms. Josephine Howe, Department of Microbiology, NUS)

flagella (Fig. 8.4). The flagella are rotated like an outboard motor, to propel them forward or backward. This helps them in coming closer to food, or moving away from noxious substances. Amazingly, bacteria are able to judge the difference between the two!

Based on their gaseous requirement bacteria can be classed into three groups.

- Aerobic   — dependent on O_2 for their survival.
- Anaerobic   — cannot tolerate any O_2.
- Facultative anaerobic — prefer growing in the presence of O_2 but can also grow without it. Most bacteria fall into this category.

Fig. 8.5. Antibiotic sensitivity test showing zones of inhibition against various antibiotics. The solid medium is first layered with the bacterium to be tested and the antibiotic disc then placed.

Bacteria lack the membrane bound nucleus of the eukaryotes and their DNA appears as a tangled web in the center, known as the nucleoid. In addition, bacteria contain plasmids, or small loops of extra chromosomal DNA; these can be transmitted from cell to cell by conjugation. Antibiotic resistance is carried by the plasmids, and can spread rapidly between bacterial species by this method (Fig. 8.4). Once resistance is acquired it is rarely lost, which means that the pool of antibiotic resistant bacteria is continuously growing, posing a danger to the human population. The antibiotic sensitivity is usually tested by the "disc diffusion" method, in which discs impregnated with antibiotics are placed on the medium layered with the organism to be tested. The antibiotics diffuse through the agar and produce a zone of inhibition. The degree of sensitivity corresponds to the diameter of inhibition. (Fig. 8.5)

In all bacteria, the plasma membrane is surrounded by a cell wall, excepting *Mycoplasma*, which are tiny organisms 0.2–0.3 μm and are pleomorphic (variable), in shape. The composition of the cell wall differs amongst species and is used for classifying bacteria into two major groups, using Gram's stain (named after a Danish bacteriologist, who developed the stain). The bacteria with cell wall made up of peptidoglycan, stain purple with crystal violet (a dye), and are designated as G+. Other bacteria with an outer wall of carbohydrates, proteins and lipids do not stain with crystal violet, and are designated as G−. Gram's staining is used very widely for diagnostic purposes. It also provides some indication of antibiotic sensitivity of the bacteria.

Many bacteria surround themselves with a thick hydrophilic gel which may be thicker than the bacterial cell. If it is well defined it is called a *capsule*, but if it is amorphous and irregular it is referred to as a *slime layer* (Fig. 8.6). Capsules protect the bacteria and impede their ingestion by phagocytes.

Bacteria can form a resistant stage or spore (Fig. 8.7). In this, a part of the genetic material of the cell gets concentrated in one place and then gets surrounded by a protective coat. This makes the spore impervious to chemical agents, dessication and temperature, which would normally kill the vegetative stage. The spore is metabolically inert and can survive for months, years and even decades. When the spores are exposed to favorable conditions, germination occurs with the emergence of the vegetative stage, which then undergoes normal replication by binary fission.

Some bacteria produce poisonous substances known as toxins. Toxins can be divided into two categories. Those that can easily be separated from the bacterial cell, are known as exotoxins and those that remain coupled with the bacterial cell, are known as endotoxins. The endotoxins are usually associated with G-bacteria. Exotoxins, on liberation from the bacteria, circulate in the blood and can damage organs distant from the site where the bacteria are primarily located. For example, in diphtheria, the causative organisms (*Corynebacterium diphtheriae*), are located in the throat, while the toxins damage the heart muscle. By chemical treatment, exotoxins can be made non-toxic, and yet retain their

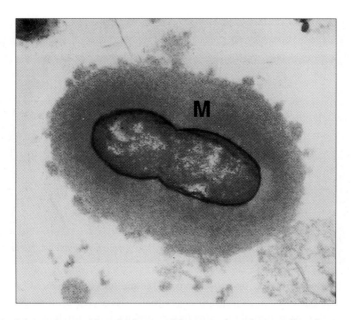

Fig. 8.6. A bacterium with a thick mucoid coat, referred to as slime layer. ×30 000. TEM. M = mucoid coat.

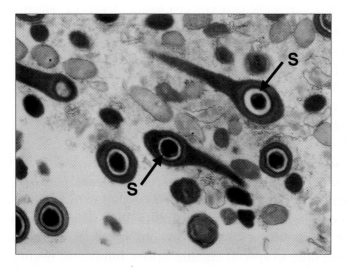

Fig. 8.7. Bacteria with a terminal spore appearing as round bodies with a thick wall. This gives "drum-stick" appearance to this species (*Clostridium tetani*). ×25 000. TEM. S = spore.

Fig. 8.8. Bacteria growing on solid media. This method is known as "streaking" in which the bacterial concentration is gradually decreased until single colonies appear. In this case hemolytic *Streptococci* are growing on blood agar. Clear areas indicate where hemolysis has occurred.

immunogenecity, when they are known as toxoid. Toxoid is used for immunization against two dangerous diseases, diphtheria and tetanus.

Most bacteria can be cultivated or grown outside the host in artificial culture media. If the medium is rendered solid by the addition of a gelling agent known as agar (a polysaccharide extracted from seaweed) the bacteria form visible masses of growth known as colonies. Colonies of different species vary in size, shape, texture and color. Colonial characters are therefore used for the identification of bacterial species. (Fig. 8.8)

When bacteria act as foes, their first step is to come in contact with the host tissues. The indigenous flora, as stated earlier, are already in contact with the host and infection can result from them, but only in special circumstances. Infections, resulting from this local pool are said to be "endogenous". An example of this is the urinary tract infection caused by *Escherichia coli*, from the intestinal tract. This occurs more often in females, because of the close proximity of the anal opening to the female urethra, and its small size. *E. coli*, which were non-pathogenic (not disease producing), in their normal habitat (intestine), become pathogenic in their new or abnormal habitat, the urinary tract. A common example of endogenous infection is dental caries. This is characterized by demineralization of the enamel and destruction of the organic matrix of teeth. The primary pathogen is *Streptococcus mutans* along with few other mouth dwelling bacteria.

Organisms acquired from external sources are labeled as "exogenous", and are responsible for many serious diseases. This kind of transmission can occur from human to human, or from animals to human. In the latter case, the infection is called zoonosis.

The major routes of transmission of exogenous infections are:

a. Direct contact (including sexual intercourse).
b. Inhalation or droplet infection.
c. Ingestion.
d. Inoculation (including arthropod bites).
e. Vertical transmission (from mother to fetus).

The body is constantly exposed to microorganisms from its environment, and these exposures, in most cases, do not result in disease. To establish themselves in a new host, the microorganisms need to survive, multiply and overcome the host defenses. If the host defenses are impaired, as in AIDS, then it becomes easier for the microbe to establish itself in the new host.

Direct contact in the form of sexual intercourse is a highly effective form of transmission. Slowly but surely the infection spreads in a community. Social and cultural behavior plays a very important role in the spread. There are three important bacterial infections which are spread sexually. These are gonorrhea (caused by *Neisseria gonorrhoeae*), Syphilis (caused by *Treponema pallidum*), and non-gonoceal urethritis (caused by *Chlamydia trachomatis*). Vaccines are not available for any of these, and there is at present no possibility of eradicating them from the human population.

Inhalation is an extremely common mode of infection and may result in the upper or lower respiratory tract involvement. Lower respiratory tract involvement is less common, but more likely to cause serious illness and even death. Transmission is more rapid in groups sharing living quarters or visiting congested places. A large number of bacteria can cause respiratory tract infections and one which has been in the limelight recently is *Legionella pneumophilia*. This bacterium is transmitted via air-conditioning ducts or by hotel shower exposure and can lead to pneumonia. *Mycobacterium tuberculosis* is also acquired, in most cases, by inhalation. In the developing world, tuberculosis is probably the most important bacterial infection. In developed countries, its incidence had decreased since the last century, with improved standard of living. The situation has now changed because of AIDS and the disease has re-emerged as major public health problem in the developed world as well. In addition, the bacteria have become resistant to many anti-tuberculosis drugs, compounding the problem of effectively treating the disease in AIDS cases.

Ingestion of contaminated water and food is particularly common in third world countries, because of poor food hygiene. If the main water source gets contaminated, major epidemics can ensue. In the past, cholera has spread in this manner and has led to the death of millions of people all over the world. The contamination of food is closely connected with house-fly population, which acts as a mechanical vector, picking up bacteria from feces and transferring it to food. Typhoid fever, caused by *Salmonella typhi*, is a very important food-borne infection in developing countries. The recovery from typhoid may be followed by carriage of the bacilli, often for many years, in the gall bladder. These symptomless carriers act as a source of further infections as bacteria are excreted in their feces.

Inoculation of bacteria by arthropods is less common than in viral infections. Tiny Gram negative bacteria, known as *Rickettsiae*, are transmitted by the bites of tick, mite, louse and flea. A variety of illnesses are produced as a result of infection by *Rickettsiae*, and at least 11 species of bacteria are involved. Historically, the most famous bacterium transmitted by the flea is *Yersinia pestis*, which is the causative organism of plague. Plague, under the name of Black death, caused death and destruction across Europe in the middle ages. Plague is still endemic in parts of Asia, but no major epidemics have occurred in the recent past.

In 1975, a new bacterial disease was recognized in rural children in Lyme, Connecticut, USA. This illness is now known as Lyme disease. The causative agent of Lyme disease is *Borrelia burgdoferi*, named after Dr. Bergdorf, the microbiologist who identified the organism. The infection is transmitted by ticks, and starts with expanding annular skin lesions around the tick bite. In 5–15% of untreated cases, cardiac, nerve and joint involvement occurs.

Inoculation of bacteria into tissues can also occur as a result of trauma and accident. The soil bacteria can enter the tissues and can cause life threatening diseases. *Clostridium tetani*, an anaerobic spore forming organism, enters the tissues in this way and causes tetanus (Fig. 8.7). The symptoms of tetanus are lock-jaw (patient is unable to open his mouth), and convulsions, which occur as a result of exotoxins secreted by the organism. In untreated cases, the mortality rate is high.

Congenital or vertical transmission of bacteria occurs in syphilis. The spirochetes are able to penetrate the placenta from the mother's blood and enter the fetal tissues. The baby may be born with a variety of defects involving bones, skin and the nervous system. Luckily, syphilis is an easily treatable disease with antibiotics, and congenital syphilis is now rarely seen.

✳✳✳✳✳

"How does a little coccus
find so many ways to knock us
flat upon our backs?

And that red or blue bacillus
find so many ways to kill us
through similar attacks?

Protozoon and virus
viciously require us
to bear their special whacks

Yeast and mould and fungus,
even worms who live among us,
have ways to take their cracks

And spirochaetes who hate us
in loving ulcerate us,
if we are rash and lax".

Samuel Stearns
From, *Perspectives in Biology and Medicine*, 1981.

9

The Mouldy World

Moulds grow everywhere and in every climate. Even glass surfaces are not spared and many camera lenses have been destroyed by moulds. Their presence is most obvious on stored bread, cheese and other food items, on which they form multicolored cottony layers.

The scientific name for mould is fungus and their study is called mycology. Fungi are eukaryotic organisms, possessing a nucleus enclosed by a nuclear membrane. Their cell wall is distinct like plants, but unlike plants, they do not possess chlorophyll and are unable to synthesize energy from sunlight. They exist as *saprophytes* (organisms that live on dead or dying matter), and cause destruction of timber, paper and many other synthetic substances. However, their role as saprophytes is also beneficial as they break down plant and animal remains, and clear the environment.

Fungi are beneficial to humans in many other ways. They are used for the fermentation of beverages, the making of cheese, yoghurt, soya sauce and other food products. They are an important source of antibiotics. The first antibiotic (penicillin), was discovered by Alexander Fleming (1881–1955) from the fungus *Penicillium*, for which he got the Nobel Prize in 1945. In 1996, the dried out culture of the original isolate, prepared by Alexander Fleming in the 1920s, was auctioned by Sotheby's of London to a pharmaceutical company for £23 000. This is of course a very small price for the value of penicillin, which saved million of lives since its discovery. In addition to antibiotics, some other important drugs are also derived from fungi. LSD comes from ergot that grows on rye and other grains. A cholesterol lowering drug lovastatin is derived from the fungus *Aspergillus terreus*. Cyclosporin A, an immunosuppressive drug used in transplantation, comes from the fungus *Tolypocladium inflatum*.

All fungi are aerobic (need O_2 for their growth), and many reproduce by both sexual (meiosis), and asexual (mitosis) division. Their size varies from 3 μm to 3 feet in diameter, e.g. mushrooms. Many fungi produce branching filaments known as hyphae. When hyphae get interwoven to form a network, it becomes a mycelium. If hyphae have cross walls, the fungus is said to be septate. The presence or absence of these cross walls is useful in identifying species which infect humans.

Fungal spores are functionally the same as plant seeds. Spores produced on the sides or tips of hyphae are called conidia. They are very light and get dispersed by air currents. The variation in spore morphology is used for speciating fungi (Figs. 9.1 and 9.2).

The yeasts are unicellular fungi. They are usually spherical or oval with a thick wall and divide by budding. In this process, a small cell forms as an outgrowth of the parent cell (Fig. 9.3). The bud gradually enlarges and separates from the mother cell. In some yeasts buds don't detach and a chain forms (Fig. 9.4). Many fungi which infect humans are dimorphic, i.e. show hyphal growth when growing saprophytically in soil or culture, and yeast forms when growing inside the tissues.

Fungi produce many human diseases and for medical purposes, mycoses can be divided into two major groups: superficial and deep. The former involves body surfaces such as the skin, hair, and nails. In view of their superficial involvement, they are also called dermatophytes and are commonly known as ringworm. There are numerous species of dermatophytes, but only three genera (*Microporum*, *Trichophyton* and *Epidermophyton*), infect humans. The most significant feature of dermatophytes is that they are all keratinophilic, which means they are attracted to the keratin of the skin and its appendages, and do not grow elsewhere

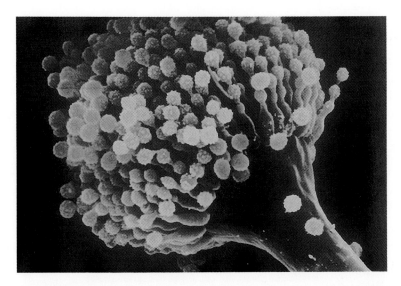

Fig. 9.1. Branched filaments or hyphae of a common fungus, *Aspergillus fumigatus*. Note the presence of conidia or spores attached to the filaments. ×2000. SEM.

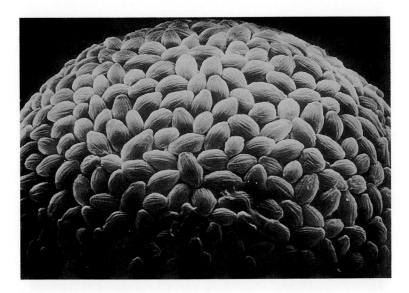

Fig. 9.2. Spores are similar to the seeds of plants with each species having a characteristic morphology. In this case the barrel shaped spores belong to a species of *Mucor*. ×4000. SEM.

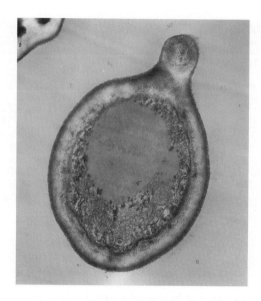

Fig. 9.3. A yeast (*Candida albicans*) in the process of budding. Note the presence of a thick wall which enables them to survive in differing environment. ×15000. TEM.

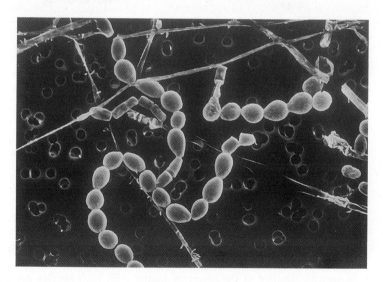

Fig. 9.4. Long chains of oval yeast cells can be seen. In this case buds have remained attached to each other. ×200. SEM.

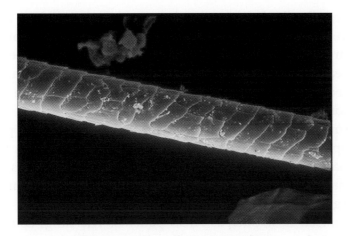

Fig. 9.5. A normal hair. ×2000. SEM.

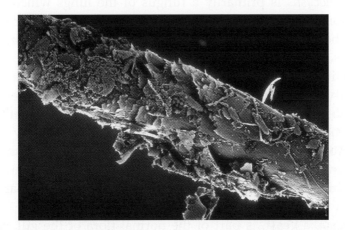

Fig. 9.6. A hair infected with fungus (*Trichophyton* sp) causing its destruction. ×2000. SEM.

in the body. Invasion of the hair leads to the destruction of the hair shaft which often breaks off (Figs. 9.5 and 9.6).

The dermatophytes are amongst the most common disease producing organisms of man and approximately 10–20% of the world's population may be infected with them. The dermatologists often describe the infection by dermatophytes according to their location, e.g. tinea capitis, infection of the scalp; tinea barbae of the beard; tinea cruris of the groin; tinea corporis of the body; and tinea pedis of the foot. Tinea pedis is also known as athlete's foot, because it is commonly seen in the athletes due to the collection of perspiration on their feet.

Dermatophytes may be passed from person to person (anthropophilic); from soil to human (geophilic); and from animal to human (zoophilic). Animal sources include cats, dogs, cattle and small rodents. Fungi which cause deep mycoses can infect any organ of the body and are especially dangerous in the compromised host, such as AIDS patients. The fungi commonly involved belong to the genera *Pneumocystis*, *Candida*, *Aspergillus* and *Cryptococcus*.

Pneumocystis is primarily a fungus of the lungs which causes pneumonia, although it may occasionally extend to other organs of the body. It is the most common fungal infection in AIDS patients. Patients usually present with breathlessness, a non-productive cough and fever. Within the lungs there are foamy intra-alveolar exudates, along with plasma cell interstitial pneumonia. Infection may continue for weeks or months but occasionally it progresses rapidly and the patient may die of respiratory failure in less than 7 days. The mode of transmission of *P. carinii* is not clear. The parasite is common in rodents and other animals but immunocompromised individuals probably get infected from human carriers.

Candida, a yeast, is part of the normal flora of the gut, mouth and vagina. Moisture, warmth and broken skin encourage *Candida* to grow and cause infection of the skin and of the mucous membrane which is commonly known as thrush. The skin surfaces which are in close proximity to each other and which trap moisture are usually involved, such as the area behind the breasts of women, the abdominal skin folds of obese persons and babies, the angle of the mouth and the foreskin of the penis. In the compromised host, *Candida* can also spread to various internal organs of the body, via the blood stream, producing serious disease.

Aspergillus is the most common fungus of the environment and we are constantly inhaling its spores. In asthmatic individuals, this can be dangerous, and induce an attack of the disease. *Aspergillus* can sometimes invade cavities in the lung which are produced by tuberculosis bacteria and form what is known as the "Fungal ball". In the compromised host, *Aspergillus* can spread from the lungs to the brain, kidney, liver and other organs of the body.

Cryptococcus is a yeast and is found in pigeon droppings. It mainly infects the nervous system and causes meningitis. Both normal and compromised individuals may be involved, but the disease is much more severe in the latter group. A species of *Cryptococcus* has also been detected growing on the leaf and bark of certain eucalyptus trees, accounting for its common occurrence. It has a large mucoid capsule which differentiates it from other yeasts such as *Candida* (Fig. 9.7).

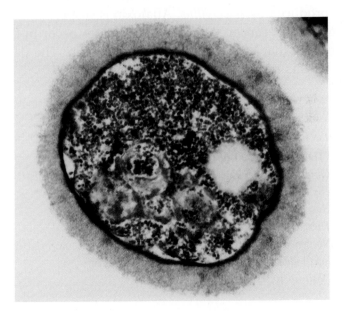

Fig. 9.7. *Cryptococcus neoformans* is also a yeast with a thick mucoid capsule. ×15000. TEM.

Superficial fungal infections are easily treatable but only a few drugs are available for the treatment of deep mycoses. For the satisfactory treatment of these infections, predisposing factors should be attended to, which are as follows:

- Individuals with impaired immunity, e.g. AIDS.
- Uncontrolled diabetes.
- Pregnancy (as the immune response is depressed during pregnancy).
- Persons with leukemia or other cancers.
- Persons taking certain medications, e.g. anti-cancer drugs, antibiotics and corticosteroids.
- Marasmic and undernourished infants.

Fungi also produce toxins and diseases produced by the ingestion of fungal toxins, including poisoning due to the eating of certain mushrooms (mycetism), and contaminated grain (mycotoxicosis). The main class of toxins on grains is aflatoxin, which is produced by *Aspergillus*. Aflatoxin is suspected of causing liver damage and cancer. Aflatoxin metabolites occur in the milk of the animals that have eaten the infected grain. The danger to humans by this route is, however, not known.

✳✳✳✳✳

"The dead are fast forgotten
They outnumber the living, but where are all their bones?
For every man alive there are a million dead,
Has their dust gone into earth that it is never seen?
There should be no air to breathe, with it so thick
No space for wind to blow, nor rain to fall;
Earth should be cloud of dust, a soil of bones,
With no room even, for our skeletons".

Sacheverell Sitwell
(Thanks to fungi and bacteria in the soil this does not happen)

10

The Unwanted Guests

Parasites are defined as organisms which live at the expense of, or at the cost of, another organism (host). This definition implies that parasites are always harmful, which is not always the case. Some parasites are very beneficial. For example, ciliates living in the rumen of the cattle produce enzymes to digest food. These parasites are called symbionts, i.e. they have a mutually advantageous relationship with the host. This relationship was probably established millions of years ago, during the course of evolution. The first mammals were apparently carnivores, with a simple digestive tract. Because of the shortage of meat, and other factors, some were forced to eat plants, but had no enzymes to break down cellulose. This was achieved when ciliates and bacteria from the soil and water made rumen their home. Without these symbionts the herbivores cannot survive.

A fascinating example of symbiotic relationship is seen in case of leaf-cutter ants. These ants grow a mushroom like fungus in underground gardens, which is their main food source. The fungus in turn is dependent on ants for its food in the form of leaves. However, the fungus can be destroyed by a green mould if it enters the garden. To prevent this from happening the ants carry a bacterium which produces an antibiotic lethal to the green mould!

Many parasites are commensals,[a] i.e. they are neither harmful nor beneficial to the host. A number of protozoa in the intestinal

---

[a]From Latin meaning "at table together".

tract fall into this group. A few parasites are certainly "unwanted guests", and cause various health problems. These are called the pathogens. A good example of a pathogen is *Plasmodium*, the malarial parasite.

Parasites are divisible into two main groups — protozoa and helminths. In a simplified classification, protozoa which are of medical importance, can be further divided into five phyla:-

- Sarcodina or amoebae.
- Mastigophora or flagellates.
- Ciliphora or ciliates.
- Apicomplexa or organisms which possess a structure known as the "apical complex" at the anterior (front) end.
- Microspora or minute tissue dwelling organisms which form spores.

The helminths are divided into three major phyla:

- Nematoda or roundworms.
- Platyhelminthes or flatworms.
- Acanthocephala or thorny-headed worms.

The platyhelminthes are further divided into two classes:

- Cestoda or tapeworms.
- Trematoda or flukes.

Unlike bacteria and viruses, parasites usually have a complex life cycle and may have one or more intermediate hosts. This makes their survival difficult, but they have devised some ingenious ways of overcoming the problem and passing from one host to another.

Parasites employ a number of methods to effect transmission. These include ingestion, direct penetration of the skin and the mucous membranes, inoculation by arthropod vectors, sexual intercourse (venereal transmission), and vertical or congenital (mother to fetus) transmission. Occasionally, parasites get transmitted by blood transfusion and transplantation of tissues.

The protozoan parasites are unicellular organisms which are widely distributed in nature and each protozoan cell is able to perform all the vital functions of life.

The following genera are amongst some of the medically important protozoan parasites:

- *Entamoeba* (an intestinal amoeba).
- *Giardia* (an intestinal flagellate).
- *Trichomonas* (a genital flagellate).
- *Leishmania* (a tissue flagellate).
- *Trypanosoma* (a blood flagellate).
- *Plasmodium* (a blood parasite).
- *Toxoplasma* (a tissue parasite).
- *Balantidium* (an intestinal ciliate).
- *Cryptosporidium* (an intestinal parasite).
- *Enterocytozoon* (an intestinal parasite).

In the genus *Entamoeba*, only one species, *E. histolytica*, is pathogenic, which is the causative agent of amoebic dysentery and is a very common infection of the third world countries. It is acquired by the ingestion of food and water contaminated by the cysts of *E. histolytica* (Fig. 10.1), and passed in the feces of the human carriers. *Giardia* is another common intestinal parasite, acquired by ingesting contaminated food and water. The parasite has a characteristic shape with a disc-like structure on its ventral surface, by which it attaches itself to the mucous membrane of the small intestine (Fig. 10.2). In heavy infection, large parts of the small intestine may be covered with the parasites, preventing absorption of nutrients, especially fats. The genus *Trichomonas* has three species, of which only one is pathogenic (*T. vaginalis*). This lives in the genital tract of the female and is transmitted by sexual intercourse. The male acts as the carrier of the parasite.

Leishmania is composed of many species, causing a variety of illnesses, ranging from the ulcers of the skin to systemic infection (visceral leishmaniasis), which can be fatal, if left untreated. The parasite is transmitted by the sand-fly. *Trypanosoma* has two species, one causing African trypanosomiasis, and the other causing South American trypanosomiasis. African trypanosomiasis is also known as sleeping sickness as it involves the brain and is transmitted by the tsetse fly. The South American trypanosomiasis is also called Chagas' disease, in which the cardiac muscle is

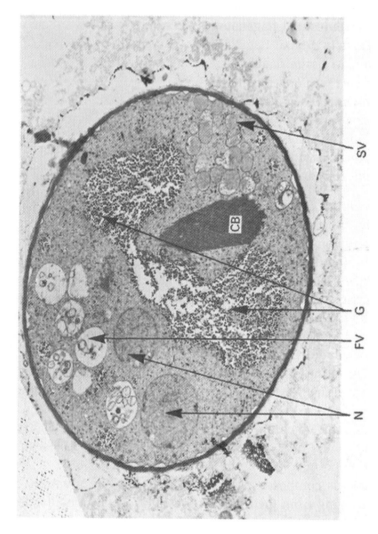

Fig. 10.1. *Entamoeba histolytica* cyst. Note the presence of distinct cyst wall which protects it from adverse environmental conditions. N = nuclei, FV = food vacuoles, G = glycogen, SV = secretory vacuoles containing enzymes, CB = chromatoid body. The chromatoid body is a complex of RNA and protein and its function is not known. ×16000. TEM.

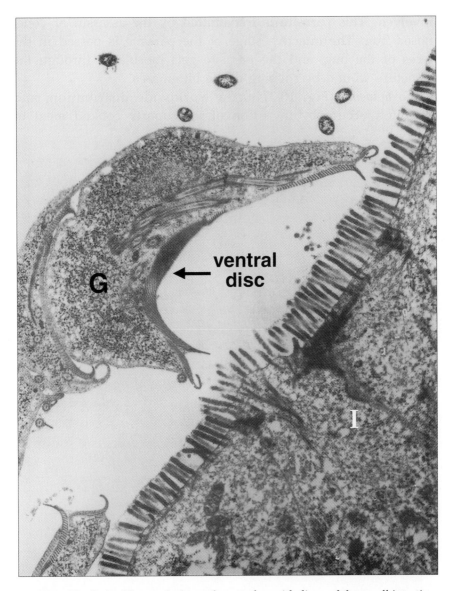

Fig. 10.2. *Giardia lamblia* attach themselves to the epithelium of the small intestine with a disc like structure on their ventral surface. The small projections on the intestine are the microvilli. In this case a *Giardia* is seen attached to the back of another *Giardia*, which is an unusual occurrence. ×30000. TEM. G = *Giardia*, I = intestine.

involved. This parasite is transmitted by the assassin or cone-nosed bug. The infective stage of the parasite is passed in the feces of the bug, and the parasite gets inoculated through the puncture wound by rubbing of the bitten area.

Toxoplasma (Fig. 10.3) has a worldwide distribution, and infection occurs by ingestion of improperly cooked meat or

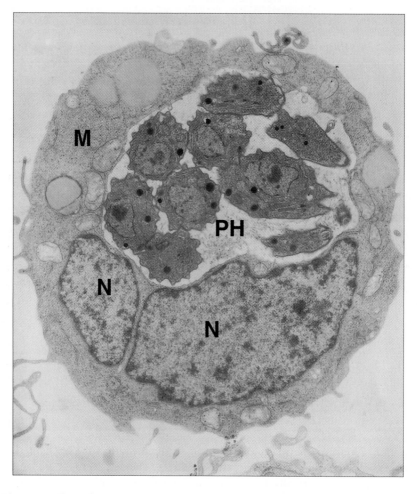

Fig. 10.3. *Toxoplasma gondii* are intracellular parasites and in this case are multiplying inside the phagosome of a macrophage. ×20 000. TEM. M = macrophage, N = nucleus, PH = phagosome.

fondling of cats, as one of the infective stages is passed in cat feces. The majority of individuals infected by this parasite are asymptomatic or show only a mild form of the disease. In the compromised host, however, infection gets reactivated and the parasites undergo unrestricted multiplication, leading to severe illness. *Toxoplasma* can also pass from the mother to the fetus and it is an important cause of abortion and congenital deformities in countries where the infection is common.

Plasmodium, which is the causative organism of malaria, has four species. One of the species, *P. falciparum*, is the most dangerous, because it can give rise to a number of complications, including brain involvement (cerebral malaria), which carries a high mortality in untreated cases. *Plasmodia* are transmitted by the bite of the female *Anopheline* mosquito and the parasite undergoes asexual development in the human, inside the red blood cells (Fig. 10.4), and sexual development in the mosquito, on its stomach (Fig. 10.5). This complicated life cycle (Fig. 10.6), and the multiple steps involved, has made it difficult to produce an effective vaccine against *Plasmodium*. Multiple drug resistance malaria is now widespread and the danger is that we may run out of treatment options, unless new drugs are discovered soon.

Balantidium (Figs. 10.7 and 10.8) is the only ciliate which infects humans. Its body is covered with hair-like cilia which enables it to move actively. It lives in the large intestine and can produce dysentery. Its main reservoir is the pig and humans get infected from pigs.

Cryptosporidium is acquired by ingestion of oocysts passed in human feces. The organism multiplies in the superficial part of the intestinal epithelium. Normally it produces little or no disease, but in AIDS patients it can produce severe protracted diarrhea.

Microspora is a large and complex group of protozoa. The genus *Enterocytozoon* is the most common microspora infecting humans and is a major cause of AIDS related diarrhea. Human-to-human transmission via the fecal-oral route is probably the most important mode of transmission.

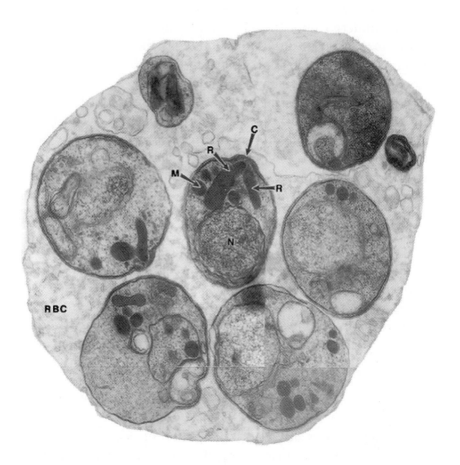

Fig. 10.4. *Plasmodium* sp. Showing a red cell containing asexual stages (merozoites) of the parasite. The merozoite has a conoid (c) at the anterior end and enzyme carrying organelles know as rhoptries (r) and micronemes (m). Each merozoite has a single nucleus (n). ×28 000. TEM.

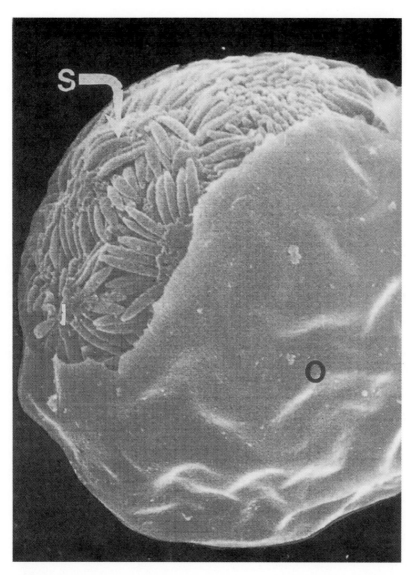

Fig. 10.5. *Plasmodium* sp. Sexual development in mosquito results in the formation of round bodies known as oocysts (o) on its stomach. Each oocyst has thousands of sporozoites (s) which are the infective stage of the parasite. ×15000. SEM.

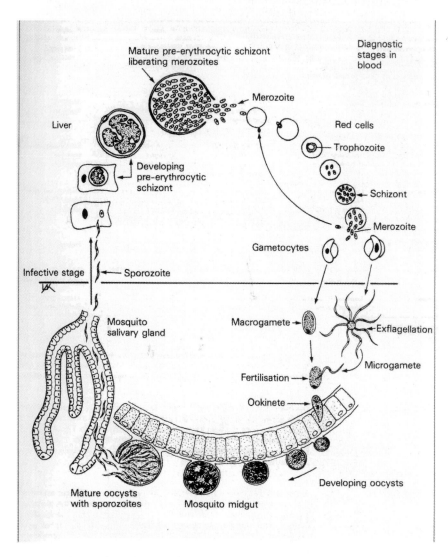

Fig. 10.6. *Plasmodium falciparum.* Life cycle showing the development stages is mosquito (lower part) and in man (upper part).

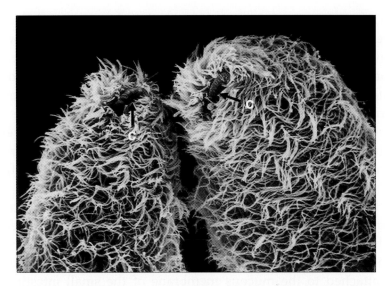

Fig. 10.7. *Balantidium coli* is a ciliate which infects man. Numerous cilia on their surface is visible. They move actively by the beating of the cilia. The slit like opening at the anterior end is the cytostome (c) or mouth. ×10000. SEM.

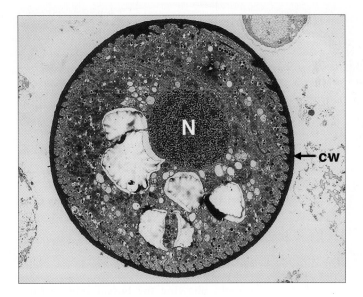

Fig. 10.8. *Balantidium coli* cyst. Note the presence of a distinct cyst wall and a large nucleus in the cytoplasm. ×8000. TEM. cw = cyst wall, N = nucleus.

The most common intestinal nematode infecting man is *Ascaris*. It resembles the earthworm in size and shape but is yellowish-white in color. It resides in the small intestine and releases approximately 250000 eggs per day. For this, it needs to obtain proteins from the host. An infected child may carry dozens of worms leading to nutritional problems. The eggs of *Ascaris* are extremely resistant to chemical treatment, and can develop even in formalin solution and can survive in the soil for months. Humans get infected by ingestion of the embryonated eggs from the soil.

Hookworms (*Ancylostoma* and *Necator*), are so-called because their anterior end is curved or hooked. This parasite also develops in the soil and the infective stage is a larva, which penetrates the skin when the individual is walking bare-footed. The adults get attached to the mucous membrane of the small intestine by their teeth or cutting plates (Fig. 10.9). After attachment they

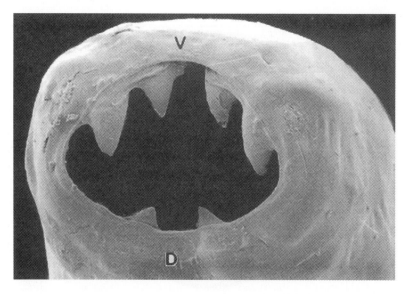

Fig. 10.9. *Ancylostoma duodenale.* Showing the buccal capsule (mouth) with 2 pair of hooks on the ventral (v) side and one pair on the dorsal (d) side. ×2000. SEM.

continuously suck blood, and an estimated 0.15 ml of blood is lost per worm, each day. Hookworms are, therefore, a very important cause of iron-deficiency anemia in the endemic regions of the world. *Strongyloides stercoralis*, like hookworms, also develops in the soil and enters human tissues by skin penetration. It is the only medically important nematode which can *multiply* in the tissues by parthenogenisis and can cause life threatening disease in AIDS patients. *Enterobius vermicularis*, also known as pinworm, is a small nematode measuring a few mm in length. The adults live in the human of the large intestine and the females emerge from the anus when the patient is at rest, laying eggs on the perianal region. This causes pruritis or itching of the anal region which can be very disturbing to children, in whom the infection is common. *Trichuris trichiura* is also known as the whipworm because of its whip-like shape. Infection occurs by the ingestion of eggs from the soil. Light infections are generally asymptomatic but heavy infections can cause dysentery and prolapse of the rectum. *Trichinella spiralis* is essentially a parasite of pigs, infecting humans through pork consumption. The larvae are carried in blood to various tissues of the body where they encyst. Heavy infections can produce serious disease and this is probably the reason why pork was prohibited in Islam and Judaism.

There are many species of filarial parasites and all are transmitted by arthropods. *Wuchereria* is the most widely prevalent genus, and causes massive enlargement of the limbs, scrotum and breasts, known as elephantiasis. This happens because the adult worms block the lymphatic drainage. Another filarial parasite, *Onchocerca*, is transmitted by the black fly, and is the most important cause of blindness in the sub-saharan savanna belt of Africa. Blindness occurs because the microfilaria (embryos), released by the adult worm, migrate to the eye and damage it.

There are three important genera of tapeworms. These are *Diphyllobothrium*, *Taenia* and *Echinococcus*. *Diphyllobothrium* is also known as the fish tapeworm, as the infection is acquired by eating improperly cooked fish. It is the largest human tapeworm and can extend up to 33 feet, producing one million eggs per day. It causes vitamin B_{12} deficiency, by absorbing as much as 80–100% of orally taken vitamin B_{12}, thus depriving the host of this vitamin, which leads to pernicious anemia.

Taenia are acquired by eating improperly cooked pork or beef. The pork tapeworm (*T. solium*), is particularly dangerous, as its larval stage (*Cysticercus celluosa*) can get into the brain and give rise to epilepsy and other serious problems.

Echinococcus is acquired by ingesting eggs, passed in the dog feces, which harbors the adult worm in its small intestine. The adult worm is attached to the dogs' intestinal tissues by scolex, which is made up of rostellum with hooks and four suckers (Fig. 10.10). The larval stage of the parasite is known as the Hydatid Cyst which develops in herbivores and humans. The liver

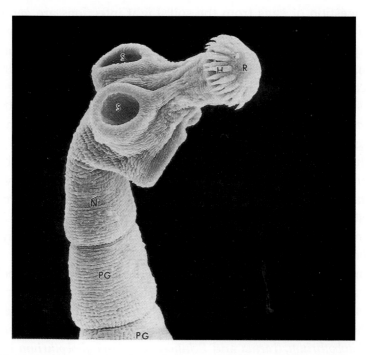

Fig. 10.10. *Echinococcus granulosus.* Scolex or head by which the adult worm attaches itself to the dog intestine. R = rostellum, H = hooks, S = sucker. ×2000. SEM.

is most often involved, followed by the lungs and brain. The cyst can grow to an enormous size and can become larger than a football. Inside the cysts are thousands of immature worms, which become adults when the infected tissues from the herbivores are eaten by dogs and other carnivores.

Amongst the flukes, two parasites are particularly important. These are *Clonorchis*, also known as the Chinese liver fluke, and *Schistosoma*, known as the blood fluke. *Clonorchis* is acquired by eating improperly cooked fish. It resides in the biliary passages of the liver, leading to cirrhosis (fibrosis) and occasionally to liver cancer.

Schistosoma are thin elongated parasites which live inside the blood vessels. They use snails as their intermediate host, and the infective stage (cercaria) penetrates the skin to initiate the infection. The adult worms remain in copulation throughout their long life of 20–30 years. The eggs, released in the blood vessels, find their way out of the body via the feces and the urine. However, a significant number are unable to do so, and get trapped in the liver, the urinary bladder and the intestinal tissues. The disease in schistosomiasis is caused by these trapped eggs. This infection was common in Egypt and calcified eggs can still be detected in Egyptian mummies.

Parasites, unlike bacteria and viruses, produce a great deal of morbidity (chronic ill health), rather than high mortality, with a few exceptions. The morbidity adversely affects the productivity of the population, thus causing economic loss. In affluent societies, many parasitic diseases have been reduced or eliminated. Similarly, affluent families and individuals in developing countries show a lower incidence of parasitic diseases, compared to economically depressed classes. Therefore, parasitic diseases are mostly diseases of poverty and ignorance.

Once poverty and disease get established, a vicious cycle develops. People become sick, because they are poor and unable to eat or treat themselves. Because they are sick they cannot work and get even poorer. One accentuates the other, debilitating and destroying whole communities.

In view of the rapid spread of international travel and movement of people across the borders, parasitic infections have seen resurgence all over the world. This is accentuated by the emergence of AIDS, which renders some relatively harmless parasites into life threatening infections. It is for this reason that parasites now command a great deal of attention even in the developed countries.

✳✳✳✳✳

"Big fleas have little fleas
Upon their backs to bite' em
And little fleas have lesser fleas
And so ad infinitum".

Jonathan Swift

11

Mad Cows and Toxic Proteins

In 1957, Carleton Gadjusek, an American physician and anthropologist, arrived in the Eastern Highlands of New Guinea to investigate a puzzling new disease called Kuru. Gadjusek found that there were three main stages in the progression of the disease. The first is the ambulant stage and includes ataxia (unsteadiness of gait), tremors, incoordination of extremities and dysarthria (slurring of speech). The second is the sedentary stage, in which the patients cannot walk, with worsening tremors and ataxia along with depression, inappropriate laughter and emotional lability. The third stage is the terminal stage in which the patient cannot sit up, faecal and urinary incontinence occurs, dysphagia (swallowing difficulty) develops and death follows.

It was obvious that this was a neuro degenerative disease and was very similar to Creutzfeldt–Jakob disease, a rare hereditary condition in the 45–75 years age groups, seen all over the world. Another scientist working on Kuru realized that the brain tissues of Kuru was similar in pathology to scrapie, a well known brain disease of sheep. Scrapie is transmissible from sheep to sheep by the injection of brain tissue, but the disease manifests only after a long incubation period of one or two years. It was, therefore, decided to inject Kuru brain tissue into chimpanzees to see what happens. In 2–3 years, the chimpanzees did develop a Kuru like illness. It was this finding that established that Kuru was an infection. But as it was not transmitted by inhalation or by physical and sexual contact, there had to be some other mode of transmission.

The inhabitants of those villages where Kuru occurred, had a very strange cultural habit. They ate human flesh in ritualistic cannibalism. The brain was particularly relished, and this indicated the route of transmission, i.e. the eating of brain. When cannibalism stopped, the disease died out, thereby establishing that cannibalism was indeed the cause of transmission. Initially, it was thought that the infectious agent was a virus, but it was not detectable by any laboratory test and could not be grown in any cell culture used for viruses. The puzzle was later solved by a neurologist, Stanley Prusiner, who put forward the idea of a prion (proteinaceous infectious particle) being the causative agent. Prion has no nucleic acid genome (all previously known microorganisms contain nucleic acid, which enable them to reproduce), so how does it reproduce?

The prion protein (PrP) is normally found on the cell surface, but in the patients its configuration changes. The abnormal protein is called PrPsc (sc for scrapie). It has been hypothesized that the distorted protein binds to the normal protein and induces it to change its configuration; this process is progressive, changing more and more proteins. This being a slow phenomenon, the disease has a long incubation period (between the beginning of infection and the manifestation of disease). In November 1986, a similar neurological disease appeared in the cattle in the UK. It was diagnosed as Bovine Spongiform Enceplopathy (BSE), also commonly known as the "Mad-Cow" disease. The name BSE was given because the infected brain looks like a sponge under the microscope, with multiple holes in it (Fig. 11.1). Initially, its cause was not known, but later investigations revealed that the disease had originated from cattle feed, prepared from the offals and blood of other dead animals, including sheep, which had died of scrapie.

On confirmation of this, the UK government in June 1988, imposed a statutory ban on feeding any material to cattle which was derived from tissues of ruminants (e.g. sheep, cattle and goats). In addition, a few million cattle, suspected of being exposed to BSE, were slaughtered to eliminate the infection from the country.

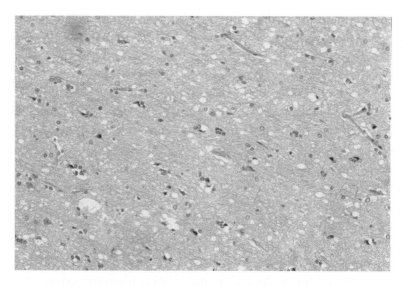

Fig. 11.1. Section of a brain of patient who died of prion disease. Note the spongy appearance with "holes" of varying sizes. ×400. LM. (Courtesy of Dr. Sheema Hassan, AKU)

However, the question which was worrying everyone at that time, was the possibility of transmission of BSE to humans, as numerous individuals had eaten beef, prior to the knowledge of its wide prevalence. Moreover, companies manufacturing the contaminated feed had already exported it to Europe and to other parts of the world, thereby spreading the infection far and wide.

Startling evidence that BSE can be transmitted to other animals, came from the work of a neuropathologist working at the Central Veterinary Laboratories, UK. He showed that BSE could be transmitted to cats, by feeding them infected beef, but his paper was prevented from publication, by the government authorities, to avoid public alarm.

Despite this crucial information, the UK government went out of its way to convince the world that its beef was perfectly safe. To prove this, a government minister fed hamburger to his four-year-old daughter on television! Roy Henderson, an epidemiologist working at the Welcome Trust, warned that it is inappropriate to underplay the possible dangers of BSE to humans. His advice was apparently ignored. In the meantime British beef was totally banned in Europe, but it appears that the exports to some other countries continued.

In 1996, the first human cases started appearing in individuals who had eaten infected beef, the disease, however, was somewhat different from the classical CJD. It was therefore given the name of a new variant CJD (nvCJD). The nvCJD affects young patients (average age 27 years versus 65 years for CJD). Psychiatric symptoms in the form of depression or schizophrenia appear early in the illness. Neurological symptoms such as convulsions and ataxia develop later. Shortly before death, the patients usually become mute and immobile. Up to September 2002, 115 deaths had been reported in the UK caused by nvCJD.

One major fear is that BSE may have also spread to sheep, which may increase the danger of transmission to humans manifold. The UK Food Standard Agency has suggested that the request to use animal derived feed even for poultry be rejected. It also suggests that a check on the sheep intestines used for sausage casings, be carried out to ensure that the processing removes lymphoid tissue, where the prions are likely to be present.

Prion diseases are not contagious and there is no possibility of direct human-to-human transmission. However, accidental transmission of prion, by contaminated surgical instruments, blood transfusion, corneal grafting and the use of cadaveric pituitary-derived growth hormone, is a possibility.

The lack of a convential immune response and the inability to detect abnormal prion protein in blood has made laboratory diagnosis of this disease difficult. At present the most satisfactory test is the detection of PrP by immuno-histochemistry, using the lymphoid tissue of the patient, such as tonsillar material.

We do not know how many humans have been infected by eating infected beef since BSE first appeared, because the disease has a long incubation period. Only the future will show whether this was a passing episode or a tragedy of monumental proportions. One thing is certain, that humans should be more careful in feeding tissues of one animal to another, as this could lead to the appearance of new and dangerous diseases.

Creutzfeld–Jakob (CJD) and its Variants

1. *Sporadic CJD* has an incidence of 0.4 to 1 case/million/year. It affects individuals between the ages of 50–75, with peak onset between 60–65 years. It is characterized by rapid progressive dementia.
2. *Familial CJD*. This occurs mainly because of the mutation in PrP gene. Inherited prion diseases appear in many forms, and different names are given to each syndrome.
3. *Iatrogenic CJD*. This is caused by accidental transmission of the prion via surgical instruments, dental procedures, injection of hormones obtained from the pituitary gland, blood transfusion and corneal transplant.
4. *New variant CJD* (nvCJD). This results from exposure to BSE. It is transmitted mainly by eating nerve and brain tissues. The number of people harboring the prion is not known because of the long incubation period.

※※※※※

"Discovery is seeing what everybody else has seen, and thinking what nobody else has thought".

Albert Szent-Gyorgi
Hungarian biochemist

(Solving the mystery of the disease Kuru won the Nobel Prize for Carleton Gadjusek in 1976)

12

Arthropods as Agents of Disease

The phylum Arthropoda (Gk = jointed foot), has three main classes, insects, crustaceans and arachnids. As a group they are the most successful life forms on earth, with more than a million known species, and many more which have not yet been described. In terms of human health they are involved in three ways:

1. As a vector or transmitter of infection, as the *Anopheles* mosquito in relation to malaria.
2. As an intermediate host, as the crustacean *Cyclops*, in relation to a nematode, *Dracunclus medinensis* (guinea worm).
3. As a causal agent themselves, as in the disease scabies, caused by a mite, *Sarcoptes scabiei*.

All arthropods have a chitinized exoskeleton (external skeleton), which is extremely tough and highly protective, but does not allow the arthropod to grow in size. The only way to grow is to split the exoskeleton and emerge with a new one. This process is called moulting or ecdysis. The newly emergent arthropod has a soft pinkish exoskeleton, which hardens in a short time and usually becomes brown or black. The internal structure of the arthropod consists of a heamocele, which is filled with hemolymph, in which its organs float. O_2 is supplied to the tissues through a set of very fine tubes known as tracheae, which open to the exterior through the *spiracles*. The cuticular walls of the tracheae are made up of spiral thickenings to prevent them from collapsing. The nervous system consists of a ganglion, synonymous to the brain of higher animals, from which a pair of central nerve cords run through the body innervating various parts. Nerve fibres also extend to external sense organs known as sensilia. Sensilia are often delicate and symmetrical (Fig. 12.1).

Many arthropods undergo complete metamorphosis (holometabolous), during their development. The cycle in this case involves the egg, larva, pupa and adult. Each stage is morphologically distinct from the next, and only the adults transmit disease. The flies and the mosquitoes belong to this group. The remaining arthropods undergo incomplete metamorphosis (hemimetabolous), and the morphological differences between the adults and the immature stages are mainly that of size and sexual maturity. Ticks, mites and bugs belong to this group, and their immature stages also transmit disease.

Fig. 12.1. Sensilia on the surface of mite cuticle. ×5000. SEM.

The two most important classes of arthropods, in medical terms, are the insects and the arachnids. The insects may have only one pair of wings (Diptera), two pairs of wings (Hemiptera) or no wings (Siphonaptera). All insects have compound eyes, which are made up of many individual units. The units are named ommatidia, and each functions as a separate visual receiver. There may be thousands of ommatidia in each eye, enabling the insects to have a very wide field of vision (Figs. 12.2 and 12.3).

The insect body is divided into three distinct regions of the head, thorax and abdomen. The adults have three pairs of legs and two sensory organs, known as antennae, which project out of the head. The arachnids, which include ticks, mites, spiders and scorpions, are not divided into distinct regions and the head is fused with the thorax. The adults have four pairs of legs, and are without wings, antenna or compound eyes.

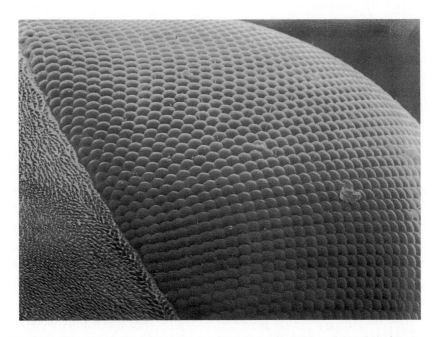

Fig. 12.2. The compound eye of the common housefly (*Musca domestica*) showing numerous ommatidia. ×4000. SEM.

Fig. 12.3. The ommatidia of *Musca domestica* at higher magnification. Note the symmetry of the arrangement. ×14000. SEM.

The medically important groups of arthropods are as follows:

MOSQUITOES

These belong to the family Culicidae. This is a very large family containing approximately 31 genera and many hundreds of species. The genera of the greatest medical importance are *Anopheles*, *Culex*, *Aedes* and *Mansonia*. Mosquitoes have an elongated mouth part or proboscis, which is a complex structure and is made up of a pair of cutting organs (mandibles and maxillae), and a sucking tube called hypopharynx, all enclosed in a protective sheath (labium). During feeding the labium bends and the hypopharynx is pushed into the tissues.

The mosquito head is globular in shape and has a pair of large compound eyes. The eggs are laid on water or on a moist substrate and are characterized by the presence of "floats" in the case of *Anopheles*. The *Culex* eggs are laid cemented together in rafts so that they can float on water. The *Aedes* eggs are often laid along the muddy edges of receding pools or in other small collections of water. They have a tuberculated surface and can withstand desiccation for years (Fig. 12.4). The *Mansonia* eggs are laid under the leaves of some water plants. The larvae and pupae of *Mansonia* obtain their oxygen by piercing these plants with chitinous siphons. This is a unique way of obtaining O_2, but it makes them totally dependent on these plants for survival.

The adult mosquitoes emerge from the pupae and mate when they are 1 to 2 days old. The female generally takes a blood meal every 4 to 5 days, after which she lays her eggs. Human *Plasmodia* (malaria causing parasites), are transmitted only by the genus, *Anopheles*. There are about 400 species in this genus, but only a few species are important as malaria vectors. Proper identification of vectors is necessary to initiate control measures, as their habitats and breeding sites differ. Mosquitoes are also vectors of filarial (nematode), and many viral infections including

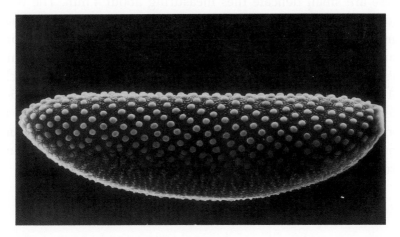

Fig. 12.4. Mosquito (*Aedes aegypti*) egg. It is characteristically ornamented and is able to withstand dessication. ×800. SEM.

dengue fever, yellow fever, and West Nile fever, which has now spread to the US from its original home in Africa and the Middle East.

HOUSE FLIES

There are many species in the housefly family, amongst which the common housefly (*Musca domestica*), is the most important. These are medium sized flies with a mouse-gray color. The head has a pair of very large compound eyes. The terminal part of the proboscis is covered with transverse ridges (pseudotracheae), through which the food is sucked in. The egg laying starts from 4–8 days after copulation and is done on a substrate which is suitable for larval development. The most common sites are putrifying organic matter and manure. House flies are mechanical carriers of bacteria and other microbes, which adhere to their feet and body, and are thus transmitted to different locations.

SAND FLIES

These are small delicate flies measuring about 4 mm. The wings and body are heavily covered with scales. The wings have numerous parallel longitudinal veins and the proboscis is shorter than in mosquitoes. The legs are thin and elongated. The female lays her eggs in cracks and holes in the ground. The larva hatches out in 1–2 weeks. It is covered with characteristic "matchstick" like hairs. Pupation occurs for about 1–2 weeks. The adults are poor fliers with a restricted flight range. Only the female sucks blood. Sandflies are vectors of the protozoan *Leishmania*, and of a virus which produces "sandfly fever".

BLACK FLIES

These are small flies measuring about 5 mm with a humped appearance. Only the female sucks blood and lays its eggs below the surface of streams and rivers. Black flies transmit a filarial worm, *Onchocerca volvulus*, which in Africa, causes a disease known as "river blindness", as transmission usually occurs in the vicinity of rivers.

TSETSE FLIES

These are robust, brown to black flies measuring about 6 mm. Both sexes suck blood and they are easily recognized by a conspicuous proboscis which projects anteriorly. The tsetse larva is fully grown in the uterus before it leaves the mother. Larviposition occurs under vegetation, in caves or in tree holes. The larva then burrows in the soil or debris and becomes a characteristic barrel-shaped pupa with distinct lobes at one end, which contain small pores through which the developing fly breathes. The adult fly emerges about one month after pupation. Tsetse flies transmit African trypanosomiasis, which is also known as the sleeping sickness. The causative organism belongs to the *Trypanosoma brucei* complex.

FLEAS

These are laterally flattened insects, without wings and measuring about 4 mm. The antennae lie in a groove in the head region. The legs are long and muscular, allowing the fleas to jump almost 200 times its own body length (Figs. 12.5 and 12.6). The eggs are laid in burrows and the larvae are maggot-like. The adults feed periodically and do not stay on their host permanently. The fleas transmit a bacterium *Yersinia pestis*, which causes plague. On ingestion of *Yersinia* by the flea, the bacteria multiply in such large numbers in the stomach that the esophagus becomes blocked. Such a flea is known as a "blocked" flea and it moves from person to person in an effort to obtain a satisfying blood meal. This causes a rapid transmission of infection.

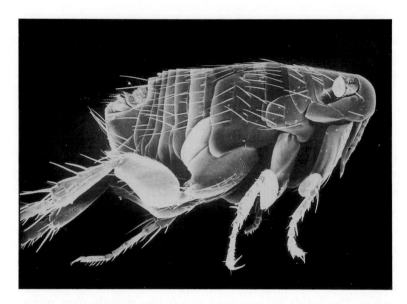

Fig. 12.5. Flea (*Xenopsylla* sp). Note the presence of the muscular hind legs which enables it to jump. ×40. SEM.

Fig. 12.6. Flea (*Xenosylla* sp) seen from above. It's streamlined shape enables it to negotiate between body hairs and clothing. ×60. SEM.

BUGS

Two families are involved, Cimicidae (bed bugs), and Reduviidae (Assasin or cone-nosed bugs). Cimicidae are wingless and Reduviidae have wings. The bodies of bed bugs are flattened dorso-ventrally and are covered by bristles. Their proboscis is short and lies folded on the ventral surface of the head. The fertilized female lays eggs after a blood meal. The eggs take 7–9 days to hatch and are operculated with a mosaic surface pattern. They pass through five moults, and all the nymphal stages are hematophagous. The bite of bed bugs can be a source of severe skin irritation and discomfort. In nature, they are not a vector of any known disease.

Assasin bugs, on the other hand, are the vectors of Chagas' disease in South America. They lay operculated eggs and have a long proboscis which is thrust forward during feeding. They are also known as kissing bugs, because they frequently bite the faces of people during sleep. In addition to their getting infected by taking a blood meal from an infected animal host, they sometimes feed on each other and infection can then be transmitted from bug to bug. The causative organism of Chagas' disease is *Trypanosoma cruzi* which is passed in the feces of the bug and enters human tissues through the bite wound left by the bug by contamination.

LICE

Lice are found in three different parts of the body, i.e. the head, body and pubic area and in each location a separate species is involved. Adult lice are 3–4 mm long, varying from grey to black in color. They attach themselves to the host tissue by strong crab-like claws and suck blood through a stylet. In the case of head lice, the eggs (nits), are firmly cemented to the base of the hair. Lice eggs are operculated and have characteristic disc-like structures with holes for the entry of oxygen. The body louse lays its eggs on clothing and the pubic louse on the pubic hairs.

The body louse transmits three bacterial infections, epidemic typhus (*Rickettsia prowazeki*), trench fever (*R. quintana*), and relapsing fever (*Borrelia recurrentis*). Typhus was a dreaded disease in the armies of the olden times. According to microbiologist Zinsser, Napolean's retreat from Moscow "was started by a louse". In the First World War, over three million Russians died of typhus. Similarly, typhus has decimated the armies of many Asian countries. In view of its medical importance, Lenin had said, "Either socialism will defeat the louse or the louse will defeat socialism".

TICKS

These are of two kinds: the Argasidae or soft ticks, and the Ixodidae or hard ticks. The capitulum or the head bears a conspicuous organ known as the hypostome, which has large recurved teeth. This allows penetration of, and firm attachment to, the host tissue. Ticks usually hide in cracks and crevices during the day. The female produces eggs in very large numbers, as the chances of survival are low. The larva which emerges has three legs. This becomes a nymph with four legs after moulting. The nymph then moults again to become an adult. In some species, there are many nymphal stages, and all are blood sucking. The ticks are a vector of many rickettsial and viral infections.

MITES

These are usually small to microscopic in size. The cephalothorax and the abdomen are fused without any line of demarcation. Four major groups are of medical importance: (a) Trombiculid mites; (b) Sarcoptic mites; (c) House dust mites; and (d) Follicle mites.

TROMBICULID MITES

These live in the soil and vegetation. Only the *larval* stage attacks animals or humans. The larvae feed on tissue fluids and lymph by embedding their mouthparts in the skin. After feeding, they drop off to moult into nymphs and then into adults. The rickettsial organisms which are carried by these mites are passed through the egg from one generation to another. This process is known as transovarial transmission. Larval mites are the vector for scrub typhus (*Rickettsia tsutsugamushi*), which is common in South East Asia and the Far East.

SARCOPTIC MITES

These include a species *Sarcoptes scabiei*, which produces the skin disease called scabies. The female mite excavates a burrow in the stratum corneum of the skin and lays about 10–25 eggs in each burrow (Figs. 12.7 and 12.8). The larvae emerge from the eggs after 3–4 days and move out of the burrow, mature and are ready

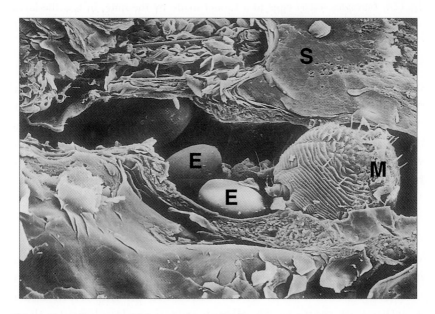

Fig. 12.7. *Sarcoptes scabiei* is a mite which causes skin disease, scabies. In this case it is seen lying in a tunnel in the skin adjacent to two eggs. ×6000. SEM. M = mite, E = egg, S = skin.

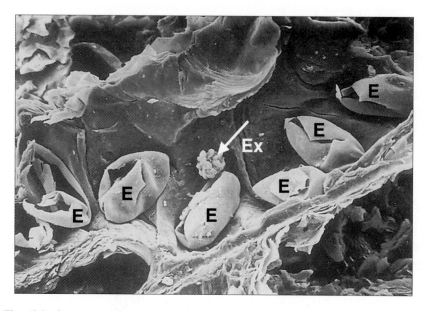

Fig. 12.8. *Sarcoptes scabiei* eggs in a tunnel made by the mite. In this the larval have hatched out and six empty egg shells are visible. ×6000. SEM. E = egg, Ex = excreta.

to repeat the process. Clinically, this condition is characterized by a papular rash, which appears all over the body, with particular involvement in the interdigital areas, axila, genitalia, buttocks, the ulna surface of the forearms and the wrist. Itching, which is often worse at night, occurs after the patient has become sensitized to the mite and its excreta. Excreta are especially allergenic.

DUST MITES

A number of species of mites are found in dust. These are microscopic and release antigenic material in the respiratory passages when inhaled. These mites are an important cause of asthma in man, as an allergic reaction develops to them and to their excretory products in hypersensitive individuals (Fig. 12.9).

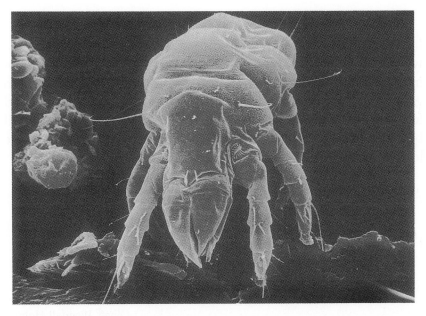

Fig. 12.9. A species of mite standing on dust particles. ×8000. SEM.

FOLLICLE MITES

These aggregate in the hair follicles and sebaceous glands. They are very tiny (0.1–0.4 mm) and are commonly located in the skin of the nose and the eyebrows. They rarely cause any medical problems.

BITING MIDGES

These are the smallest of the biting flies, measuring between 1–4 mm. The female is a blood feeder and deposits her eggs on plants or on objects close to or covered with water. The larva is worm-like with a distinct pigmented head. Medically the most important genus in this group is *Culicoides*. They are vector of a filarial worm belonging to the genus *Mansonella*.

VENOMOUS INSECTS

In addition to the commonly known venomous insects such as wasps, bees and spiders, the centipedes and scorpions are of medical importance in the tropics. In the centipede the poison glands are connected to the first pair of legs which are modified to form poison claws. In the scorpion the last segment of the tail bears the poison sac. The scorpion venom can produce severe local reaction and occasionally circulatory failure and death.

※※※※※

"Science is a good piece of furniture for a man to have in an upper chamber provided he has common sense on the ground floor".

Oliver Wendall Holmes
1870

13

The Never Ending Battle

From the day we are born, till the time we die, a never ending battle rages between the body and the invading parasite. In ancient times, the battle generally went in favor of the parasite.[a] At present, with modern weaponry such as antibiotics, the battle is usually won by the humans. For this reason, infectious diseases in the developed world are of secondary importance to the degenerative illnesses. The same is not true of the third world, where infectious diseases are still the top killers.

The host-parasite battle may progress in the following ways:

1. *Elimination of the parasite without disease*: The parasite invasion goes unnoticed by the host and the invading parasite is annihilated before it can produce any disease. For example, we are constantly inhaling *Aspergillus* (fungus) spores from the environment, which are killed by the alveolar macrophages in the lungs. Only if the host defenses are impaired will the fungus grow in the lungs. Similarly, simple procedures like brushing the teeth, send oral bacteria into the blood stream, but they are promptly removed by the white cells of the blood and normally no disease results.

2. *Elimination of the parasite following disease*: In this case the parasite invasion initially succeeds and disease follows, but

---

[a]The word parasite is used here in a broad sense and includes all microbes.

the host soon recovers and eliminates the parasite. An example of this is the common cold and a number of other minor infections.

3. *Truce between the host and the parasite*: A large number of the world's population gets infected with tuberculosis bacteria (*Mycobacterium tuberculosis*). However, the bacteria, in most cases, quietly settle down in a lymph gland or some corner of the lungs. They spring back to life when the immunity drops, as in AIDS, or in old age; otherwise the person may remain disease free all his or her life. Similarly, a species of malarial parasite (*Plasmodium vivax*), can stay dormant in the liver for many years, without giving rise to any disease. If and when suitable conditions arise, it invades the blood and produces malaria. An arthropod, *Demodex folliculum*, belonging to the mite family, lives in the hair follicles of many people, but does not give rise to any medical problem nor does it pose any future danger. Thus, the truce between the host and the parasite can be temporary or permanent.

4. *Elimination of the host*: This is normally caused by highly virulent organisms, which are able to overcome the host defenses, or it may happen if the host defenses are already impaired and are unable to resist even less virulent microbes. But ironically, this is not always the best outcome for the parasite, as in the process the parasite may also die.

The host has *specific* and *non-specific* defenses against the invading parasites. Amongst the non-specific defenses, the skin is the most important. The outer layer of the skin (epidermis), is impermeable to most microbes and is made up of dead cells, which are continuously being shed, thus discarding the bacteria that land on it. To enter into the deeper tissues, the bacteria have to break through the epidermis, as may happen in an injury. The skin is, however, not an effective barrier against all attacks. Hookworm (a nematode) larvae, and *Schistosoma* (a trematode) cercariae, can penetrate the unbroken skin. Arthropods, like mosquitoes, are also able to breach the skin with their biting mouth parts and inject parasites.

The skin has its own bacterial flora, which is inhibitory to foreign organisms, thus playing an important protective role. The salt content and acidic pH of the sweat also inhibits many exogenous microbes. In addition, sweat contains lysozyme, which dissolves the cell wall of certain bacteria. The lysozyme content of the tears is especially high, protecting the delicate epithelium of the conjunctiva and the cornea.

There are three organs of the body which are unprotected by the skin, as they open up to the exterior. These are the respiratory, intestinal, and genito-urinary tracts. The body has, therefore, developed special defenses to guard all the three systems.

a. *Respiratory tract*: Several defenses exist in the respiratory tract starting with nasal hairs, which filter coarse particles from the incoming air. A highly efficient "mucociliary escalator" consisting of ciliated epithelium and mucus, lines the trachea, and continuously propels all the inhaled particles towards the nose and mouth. The sneeze and the cough reflex also help in expelling any foreign matter. Small particles, in the size range of 5–10 μm, may pass through these barriers and reach the alveoli, where they get attacked by the alveolar macrophages (large phagocytic cells). If all these defenses fail, inflammation sets in, and neutrophils (white blood cells) pour in from the blood. At this stage, the host is sick, with fever, because inflammatory cells produce cytokines (peptides), which raise the body temperature. Fever itself is inhibitory to the multiplication of bacteria and to try and reduce it (unless it is too high), is counter productive to the body's effort to fight the infection. The specific defenses now come into play, the details of which are discussed later.

b. *Gastrointestinal tract*: The stomach acid (pH 3), kills most bacteria that are ingested. Individuals with impaired gastric secretions, therefore, run a greater risk of infection. Beyond the stomach, digestive enzymes and bile kill off many invaders. Peristaltic movements also prevent foreign bacteria from getting a foothold in the intestines. The resident bacterial flora, as in the skin, are inhibitory to the invaders. The intestine is lined with mucus producing cells (goblet cells), which secrete mucus during infection, providing a protective coating and flushing the system. Antibodies belonging to the IgA class are also secreted to help in destroying the invading organisms.

c. **Genito-urinary tract:** The vagina has a mixed bacterial flora but a bacterium, *Lactobacillus* sp predominates. *Lactobacillus* ferments glycogen producing lactic acid. This helps to maintain the low pH of the adult female vagina and prevents the growth of other microbes. The passing of urine flushes the urinary tract, and keeps it relatively free of bacteria. In the males, only the lower urethra and the foreskin has a resident flora, while the rest of the urethra is bacteria free. Secretions from the urethral and prostate glands also inhibit the bacteria. The female urethra is less likely to be sterile, because of its relative shortness and its proximity to the anus. Urinary tract infections are therefore more common in females than in males.

The *Immune System* protects the whole body against invading organisms and helps to eradicate those that have already been established. The importance of immunity is dramatically illustrated in case of the acquired immuno-deficiency syndrome (AIDS), in which the failure of immunity ultimately leads to death. Conversely, stimulating the immune response, as is done in vaccination, protects the host. The eradication of smallpox was based on this strong immune response produced by vaccination.

Immunity is divisable into *innate* and *adaptive*. Innate immunity is the first line of defense and protects the host before adaptive immunity comes into play. Innate immunity has two soluble components, i.e. complement and cytokines. The cellular components of innate immunity are the phagocytic cells and the natural killer cells (NK cells).

Complement compromises of a group of more than 20 substances which have so far been identified. Complement gets activated when a foreign substance, such as a microbe enters the body, setting a cascade phenomenon in which the product of one reaction acts as a catalyst to the next. The complement components are designated by the letter "C" followed by a number. C3 is the most important, and its cleavage forms the basis of all complement mediated phenomenon. Complement helps in the lysis of microbes, augmentation of antibody response and chemotaxsis (attracting) white cells.

Cytokines are small proteins with molecular weight ranging from 8000 to 40000 daltons. Nearly all cells are capable of synthesizing them. Some cytokines induce inflammation, while others suppress inflammation so that the body returns to normal once the invasion is over. Many cytokines have been identified and it turns out to be a highly complex system. One of the cytokines known as interferon plays an important role in protecting the cell from viruses.

The main phagocytic cells of the body are macrophages and neutrophils. Both have enzymes containing vacuoles called lysosomes, which fuse with the newly formed phagocytic vesicle or phagosome, exposing the phagocytosed organism to a variety of proteolytic enzymes. Some macrophages remain fixed in the internal organs, such as the Kupffer cells of the liver, while others circulate and are known as wandering macrophages (Fig. 13.1). The circulating phagocytes are like policemen on the beat, picking up invaders who have breached the natural defenses and have entered the body. The natural killer cells (NK cells) are lymphocytes because they can destroy virally infected cells and tumor cells without preactivation, thus providing an immediate defense and protection.

Adaptive or *acquired immunity* is a specific defense system of the host. It has two components: humoral and cell mediated. Both work together to provide a highly effective defense. Humoral immunity involves the production of immunoglobulins (proteins), which circulate in the blood stream and bind specifically to foreign substances, such as bacteria and toxins. Binding to toxins neutralizes them and binding to the microbes blocks their ability to attach to the receptor sites on the host cell. The binding also marks the microbes, so that they get picked up by the phagocytic cells, a process known as opsonization.

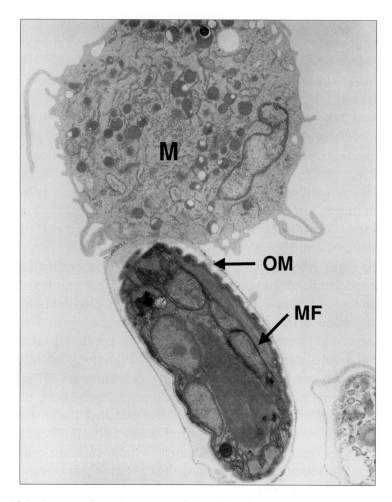

Fig. 13.1. A macrophage is seen attacking a blood parasite (microfilaria) and it has chewed a part of its outer membrane. A large organism such as this cannot be ingested by a single macrophage. To accomplish this many macrophages may join up to form a "giant cell". ×40000. TEM. M = macrophage, MF = microfilaria, OM = outer membrane.

The speed with which the immune system responds depends on whether or not the host has been primed by previous exposure to the relevant microbe, through natural infection or through vaccination. If exposure has occurred previously, the response is rapid like an army which has been trained to fight a particular enemy at a short notice.

Lymphocytes are the main cells involved in adaptive immunity and are responsible for both the humoral and cellular components. The cells of the humoral immunity are called B cells, and those of the cellular immunity are known as T cells. The term B cell is used because these cells mature in the bone marrow in mammals and the bursa of Fabricius in birds. The T cells are so-called because they mature in the thymus gland. Both B and T cells are concentrated in the lymphoid tissues of the body, e.g. lymph gland, spleen, appendix, tonsils and peyer's patches of the small intestine.

T cells circulate in the blood and lymph, attacking the cells that are harboring any microbes (Cytotoxic T cells). They help the phagocytic cells and the production of antibodies by the B cells (Helper T cells). T cells are also important in the formation of granuloma, a defensive wall which the body builds in order to cordon off organisms.

Each B cell has a specific molecular marker sticking on its surface. As there are billions of B cells in the blood, they have the ability to bind chemicals of any molecular profile in the environment. By a process known as clonal selection, B cells, recognizing foreign antigens, are stimulated to multiply and differentiate into many antibody producing plasma cells, and memory B cells, which are capable of responding quickly to any subsequent encounter with the antigen.

For the parasite to survive and infect the host, it has to first overcome non-specific, and then the specific defenses. Every parasite has developed its own unique way of doing this. Protozoa, which are destined to enter the gastrointestinal tract, enclose themselves in a cyst, whose wall remains unaffected by the stomach acid. Similarly, helminth eggs, like *Ascaris*, easily pass intact through the stomach because of their thick egg shell. Bacteria and viruses may die during the process, but if they are ingested in sufficiently large numbers, some will survive and multiply rapidly after passing through the stomach.

Trypanosomes, which cause sleeping sickness in Africa, circulate in the blood. The host responds by producing antibodies, but by the time the antibodies gain control and eliminate the parasite, it changes its outer coat. The host then produces new antibodies and the parasite responds by again changing its coat. This process continues with the parasite remaining ahead of the antibody production. It has been revealed that a hundred or more genes are involved in coat changes of the Trypanosome. Each gene is responsible for one kind of coat molecule and only one gene is active at a time. Many parasites also change their location during their development cycle; for example, *Ascaris* larva, after being released from the ingested egg, migrates from the intestine to the liver, from the liver to the lung, and from the lung back to the intestine. The net result is that the antibody response may be against antigens (foreign proteins) that are no longer relevant, or it may be directed at the wrong places, where the parasite no longer exists.

Malaria parasites (*Plasmodium* spp), pose an equally difficult problem for the body to handle, because they also undergo many antigenic changes in the blood, and remain mostly locked inside the red-cells, which antibodies cannot penetrate. A species of *Plasmodium, P. falciparum*, has an additional trick for its survival. It has surface proteins which protrude out as "knobs". With the help of these knobs it adheres to the endothelial lining of the blood vessels, thereby avoiding being swept into the spleen, where it could get destroyed. *Toxoplasma*, another protozoan parasite, can survive in the body for many years by producing a cyst wall around itself. The cyst wall does not evoke any inflammatory reaction, so the body does not recognize it "as non-self" and leaves it alone. A nematode, *Trichinella*, can also live inside the muscles of the host for many years, by making a thick wall around itself, which is impermeable to both antibodies and T cells. Trophozoites of *Toxoplasma* (see Fig. 10.3), survive inside the macrophage by preventing phagosome-lysosome fusion, and thus preventing their destruction, by the host enzymes. The same escape mechanism is used by the tuberculosis and leprosy causing *Mycobacteria*, which also manage to survive and multiply inside the macrophage. *Listeria,* another intracellular bacterium, uses a slightly different strategy; it lyses the membrane-bound phagosome and escapes into the cytoplasm of the macrophage. Many extracellular bacteria such as *Klebsiella, Hemophilus* and *Bacillus,* have a polysaccharide capsule or a slime layer around them which impedes their ingestion by neutrophils and macrophages. *Neisseria gonorrhoeae,*

Antibodies

Antibodies are specific proteins of the globulin class known as immunoglobulins. There are five types of immunoglobulins, IgG, IgM, IgA, IgE and IgD. The letters Ig stands for immunoglobulin.

1. IgG is the major immunoglobulin of the body. Its most significant role is the neutralization of soluble toxins, e.g. exotoxins and the viruses that enter the blood. It is involved in opsonization of foreign particles by enhancing phagocytosis. IgG is actively transported across the placenta and confers passive immunity to the fetus.
2. IgM mediates complement-directed cytolytic activity of foreign cells, such as bacteria. It is the first antibody to appear after an antigenic stimulus and is very good at agglutinating microorganisms. It does not cross the placental barrier.
3. IgA provides local immunity at the mucosal surfaces by preventing the adhesion of microbes to the host cells. It also occurs in breast milk.
4. IgE binds to mast cells and basophils, and triggers histamine release when antigens get attached to these cells. Histamine causes dilation and increased permeability of the blood vessels, thereby allowing the white cells to enter the inflammatory site and overcome the infection. Stimulated mast cells also secrete factors attracting eosinophils in parasitic infections. The role of esinophils is not clear, but they may be producing substances toxic to the parasite.
5. IgD is present in very low concentration in blood. Its function is not fully understood. It may serve as a recognition site for antigens by binding onto some activated B cells.

the causative organism of gonorrhea, has developed pili or hair-like structures on its surface, so that it adheres firmly to the urethra and is not flushed out during urination. *Legionella*, a bacterium which causes pneumonia in humans, multiplies inside a free-living amoeba (*Acanthamoeba*), and is protected by the amoeba's cyst wall. This very interesting "Trojan-horse" strategy, enables it to survive in water and cause infection by inhalation of the water droplets. Once inside the body, it comes out of the *Acanthamoeba*, and invades the lung. These are only some examples of the strategies which the parasites use to circumvent the host defenses. In short, the parasites have not lagged behind their hosts in the struggle for existence, and the battle between the two continues, as it has for millions of years, with no clear winners. The infectious diseases have therefore been described as "an inconclusive negotiation" between microbe and human.

✳✳✳✳✳

"The most beautiful experience we can have is the mysterious. It is the source of all true art and science. He to whom this emotion is a stranger, who can no longer pause to wonder and stand rapt in awe, is as good as dead, his eyes are closed".

Albert Einstein

14

Bolstering Defenses

Smallpox has ravaged human societies all over the world for at least 3000 years. Historians think that the disease first emerged in Egypt. The mummified remains of Rameses V (1160 BC), are suggestive of having smallpox. The disease probably spread from Africa to Asia via the trade routes.

In India the people were so terrified of smallpox that they created a goddess, Sita Saptami, in its name. The goddess was worshipped to give protection whenever an outbreak of the disease occurred.

It is thought that Arab expeditions carried smallpox to Europe, during the 6th century, where it became endemic. The term *smallpox* was used to distinguish it from *greatpox* or syphilitic rash, which was also common in Europe during those days. Syphilitic rash develops all over the body, but it does not leave deep scars as does smallpox.

In 1788, an epidemic of smallpox broke out in Gloucestershire, England, and during this outbreak a country doctor by the name of Edward Jenner observed that those patients who worked with cattle and had come in contact with a similar disease called cowpox, rarely got smallpox. He felt strongly that exposure to cowpox gave them immunity against smallpox.

On 14 May 1796, he got the opportunity to test his hypothesis. A young milkmaid by the name of Sarah Nelmes came to see him with blisters on her hand. Jenner immediately realized that she has cowpox and that this was an excellent opportunity to find out if cowpox does prevent smallpox from developing. To test this, he required a volunteer and chose his son James for this purpose. Jenner made two small cuts on James's left arm and then applied the liquid extracted from Sarah's cowpox blisters. He bandaged the wound and waited to see what happened. James developed a local reaction to cowpox, but soon recovered.

Jenner waited for six weeks and then vaccinated James again, using the same technique, but with an aspirate from a smallpox case. This experiment was extremely hazardous because if James had come down with smallpox and died, Jenner would be regarded as the murderer of his own son. To Jenner's delight and relief, James did not get smallpox.

Jenner had opened a new pathway to prevention, not only of smallpox, but many other infectious diseases. The name "vaccination" was given to the procedure from the Latin word, *Vacca*, meaning Cow. Jenner's experiment succeeded because of *cross* immunity between cowpox and smallpox producing viruses.

Eighty years later, French chemist Louis Pasteur (1822–1895) made a lucky discovery which revolutionized the science of vaccination. In 1879 while working on chicken cholera caused by *Pasteurella antiseptica*, he found that a culture that was kept over the summer months had lost its capacity to produce disease but when injected into chickens, it protected them from infection. He rightly concluded that it is possible changing a live organism that does not produce disease but still gives protective immunity; in other words, producing an *attenuated* strain of a microbe. He demonstrated the same phenomenon with rabies virus. When the virus was passaged through dozens of rabbits, it became avirulent to dogs, but retained its capacity to produce antibodies and protect the animal. His vaccine made from dried spinal cords of rabbits was then used in humans and it helped saved thousand of lives.

Immunity resulting from vaccination is based on the fact that the immune system has an extraordinary memory. In essence, each encounter with an invader, including a microbe or a foreign substance, stamps a genetic "blueprint" onto a certain number of

B and T cells. The next time these cells encounter the same foreign substance, the response is fast and powerful.

Vaccination is an *active* form of immunization, but immunity can also be conferred *passively* by the injection of the antibodies prepared elsewhere from humans or animals. Passive immunization is used to prevent or modify disease following exposure to an infective agent. Passive immunization can be combined with active immunization when the incubation period of the disease is long, as in rabies, and protection is immediately needed. Passive immunization is also used in compromised individuals who are exposed to infections which are normally benign but may become serious because of impaired defenses.

Primary Immunization Schedules*

Vaccine	Nature	Route	Age
BCG	Live bacteria	ID, SC	Birth or 10–14 years
DTP	Toxoids of diphtheria and tetanus; killed pertussis	IM	2, 3, 4 months
Hepatitis B	Surface antigen	IM	1, 6 months
MMR	Live attenuated viruses	SC	12 months
Poliovirus	Live attenuated virus	Oral	2, 3, 4 months
Hemophilus influenzae Type b (Hib) conjugate vaccine	Hib polysaccharide conjugated to protein	IM	2, 3, 4 months
Rubella	Live attenuated virus	SC	10–14 years females only (To prevent congenital transmission of the virus)

*This schedule is followed in most countries. Live attenuated virus vaccines are generally contraindicated in pregnancy.

ID = intradermal   SC = subcutaneous   IM = intramuscular

The current pediatric and adolescent immunization schedule in the USA is available on http://www.cdc.gov/nip/recs/child-schedule-jul-dec-rev.pdf

Vaccines can be of the following types:

1. *Live attenuated Virus Vaccine*: These are manufactured by attenuating the wild virus that causes the disease, so that its recipient develops immunity but not the disease. The attenuation is done by growing the wild virus through many generations until is loses its virulence. The immunity induced by this method is generally strong and long lasting. The disadvantages associated with the live virus vaccines is that it requires to be stored in a freezer, which may be difficult to obtain in the rural areas and poor countries. There is also a remote chance that the virus may mutate and then cause serious disease.

2. *Live attenuated bacterial vaccine*: The only live injectable bacterial vaccine used is BCG (named after its originators, Calmette and Guérin), which is an attenuated form of *Mycobacterium bovis*. Its efficacy in preventing tuberculosis is doubtful, but it is said to be useful in preventing tuberculus meningitis in infants. It also appears to offer some protection against leprosy. Live attenuated oral vaccine is available against typhoid (*Salmonella typhi*) and cholera (*Vibrio cholerae*).

3. *Killed vaccines*: Most of the vaccines used at present belong to this category. They have the advantage of stability even at room temperature and there is no chance of accidentally producing a virulent strain. The disadvantage is that the material contains the entire organism and it often gives rise to local and sometimes severe systemic reaction. In this regard, vaccine used against pertussis or whooping cough needs special mention, as it has occasionally caused brain damage. For this reason, the acellular pertusis vaccine which has been produced, only has immunogenic components of pertussis (components vary from 2–5) and is therefore safe.

4. *Toxoid vaccine*: In the case of a few organisms, the disease is mainly or completely due to toxins. In such cases the immunization against the toxin will not prevent the infection, but the manifestation of the disease. The material used for this purpose is called toxoid and is prepared by the chemical

treatment of toxin, so that it loses all its toxic properties, but is similar enough to the toxin to induce immunity. The two most commonly used toxoids are the diphtheria and tetanus toxoids.

5. *Subunit vaccines*: In this vaccine, the immunogenic component is separated from the whole bacterium. It is used mainly for bacteria which have an outer polysaccharide coat, which is separated and used as a vaccine. The bacteria involved are *Hemophylis influenzae, Streptococcus pneumoniae* and *Neisseria meningitides*. All these organisms produce serious diseases in children. Unfortunately, pure polysaccharide vaccines are poorly immunogenic in children of under two years of age. The polysaccharide vaccines are therefore usually bound to a protein, such as the diphtheria or tetanus toxoid, to increase its immunogenicity.

6. *Combination vaccine*: Two or more vaccines are combined and administered together to save time and expense. In this category are diphtheria/pertussis/tetanus (DPT) and measles/mumps/rubella (MMR) vaccines. There has been concern that MMR may have some connection with autism, but this has not been proven.

7. *DNA vaccines*: There is a great deal of interest in inducing protective immune response by injecting DNA sequences from infectious organisms into a carrier organism such as the vaccinia virus. Once delivered into the host, the carrier (and the inserted DNA) undergo replication, the protein of the infectious organism is produced, and the immune response develops. This technique has worked well in animals. But in humans, only DNA vaccine against hepatitis B has produced protective antibodies. The ability of DNA vaccines to deliver specific nucleotide sequences from tumor cells such as melanoma is also being explored, and holds the promise of treating this highly malignant form of cancer.

Research is in progress to incorporate microbial DNA into genetically engineered plants. Such plants would then produce vaccine in their edible parts. As these plants can be grown locally, vaccines could become available indefinitely and cheaply for the developing nations.

A novel use of vaccination has been suggested recently to break the dependence on nicotine and to help in the cessation of smoking. The principle behind this immunotherapeutic approach is to immunize the individual against nicotine. Once this is achieved, the antibodies would bind to nicotine and prevent it from entering the brain. Animal experiments show that this approach does work and human trials are awaited. Based on the same principle, a phase I trial of a human cocaine vaccine has been completed. If it succeeds, the trial could be extended to other addictive drugs as well.

Salk and Sabin

These are the two famous names in the field of polio immunization. As a result of their discoveries polio is almost on the verge of extinction.

Jonas Salk (1914–1995)

He pioneered an inactivated injectable polio vaccine, which was safe to administer and highly efficacious. He refused to patent the vaccine or to profit from his discovery. He was awarded Nobel Prize in 1954.

Albert Sabin (1906–1993)

He showed that the poliovirus entered the body via the oral route, multiplied in the intestinal tract and invaded the nervous system. He developed live but attenuated oral vaccine. It provided longer lasting immunity than Salk's vaccine and easy to administer. His oral vaccine is now used all over the world.

✳✳✳✳✳

"Today's question is not, whether there is life after death — but whether there is life after birth".

Albert Szent-Gyorgi
Hungarian biochemist

15

Microbes as Weapons

The spectre of biological warfare is haunting the world because of the anthrax outbreaks in the United States and the possibility of its usage in future terrorist attacks. Biological weapons, unlike the mustard gas used in the First World War, are directed mainly against the civilian population and in reality have little military value.

The World Health Organization (WHO), defines biological weapons as those "that depend for their effects on multiplication within the target organism, and are intended for use in war to cause disease or death in men, animals, or plants". In addition, weapons that involve microbial toxins also fall into this category, as some toxins are extremely poisonous. For example the botulinum toxin, derived from the bacterium *Clostridium botulinum*, is estimated to be three million times more potent than Sarin, the nerve gas, which was used by the Japanese cult, Aum Shinrikyo, on the Tokyo subway in 1995, killing 12 people and hospitalizing five thousand. However, if toxins are made synthetically then they fall in the category of chemical weapons.

Most microbes suitable for biological warfare can be grown quite easily, but pose a serious containment problem if and when mass-produced. Being invisible and odorless, they can spread into the general population and kill one's own citizens. Weaponization of any agent which can get easily dispersed in the atmosphere would also require considerable technical capability. Nevertheless, the idea of biological warfare is not new, and there are records of warring armies using it from ancient times.

In the 14th century, the Crimean port of Kaffa on the Black Sea, was besieged by Janibeg, Khan of the Kipchak Tartars. Some Genoese traders who had come to Kaffa also got trapped. The siege lasted for three years during which plague (caused by the bacterium *Yersinia pestis*), broke out in Central Asia and spread to Crimea causing many deaths in the Tartar army. Janibeg was forced to lift the siege and withdraw his surviving troops. As a parting shot, he catapulted the plague-infected corpses into the besieged city, knowing that the disease was contagious and would spread rapidly. The fleas, which carry *Y. pestis*, apparently transmitted the bacterium from the dead bodies to other individuals and plague broke out in Kaffa. The Genoese traders traveling by ship carried the infection back to their homes in Italy. From Italy the plague spread all over Europe and that was the beginning of the so-called "black death", which killed 25 million people or almost half the population of Europe. However, this was not the only introduction of *Y. pestis* into Europe, but it may have been the first that led to a pandemic of this magnitude.

A very devious act of biological warfare was conducted by the British forces against some Red Indian tribes of North America in the 1750s. Red Indians had no immunity to smallpox, as the disease did not occur in the Americas, before the advent of the white man. Knowing this, the British distributed blankets contaminated with smallpox scabs to the Red Indian tribes who were siding with the French. In the ensuing epidemic, 30% of the Red Indian population perished and many became blind, this being a major complication of smallpox.

In modern times, Japan was the first country to develop a biological warfare program. By 1939, they had established a biological warfare research and production center known as Pingfan near the city of Harbin in North Eastern China and in Mukden in Manchuria. Pingfan had produced a number of biological weapons, using the agents of anthrax, cholera, typhoid and bacillary dysentery. The Chinese claimed that in 1944, Japan air-dropped biological weapons on 11 Chinese cities which led to the outbreak of plague, killing many people. In a historical ruling, a Tokyo court acknowledged for the first time on August 27, 2002, that the unit 731 of the Japanese army, had used biological weapons against the Chinese during the Second World War.

The British also had a very well developed biological warfare program during the Second World War. Experiments were conducted mainly with anthrax, a potentially lethal disease, which infects both animals and humans and is caused by *Bacillus anthracis*. The site of these experiments was Gruinard Island, off the coast of Scotland. This island remained uninhabitable after the experiment for over forty years, as the spores (resistant forms) of *B. anthracis* can remain viable for decades (Fig. 15.1). Apparently, Winston Churchill, in response to German V-1 rocket attacks on London, considered bombing German cities with anthrax bombs. If this had happened perhaps 50% of the population in the cities would have perished and the cities rendered uninhabitable for decades. The danger of such an action was possible German retaliation, causing a similar destruction of British population. This factor probably prevented the use of biological weapons during the Second World War.

The Cuban government claims that it has been subjected to biological warfare by the US on many occasions. In 1971, the country experienced the first serious outbreak of Swine flu in the Western Hemisphere, which devastated its economy, as 500 000 pigs had to be killed to arrest the outbreak. Cuba also claims that microbes against sugar cane and tobacco plants have been introduced into the country from time to time to destroy these plantations. None of these allegations have ever been proved, but a more compelling case of biological warfare can be made against the white minority regime of Rhodesia (now Zimbabwe). The largest recorded outbreak of 10 000 cases of anthrax occurred between 1979 and 1985, in the territory controlled by the guerillas fighting the government. It is unlikely that such a large number of individuals could have been infected by any natural means. No serious investigations were made of this incident because of political reasons and it was attributed to the breakdown of veterinary public health practices.

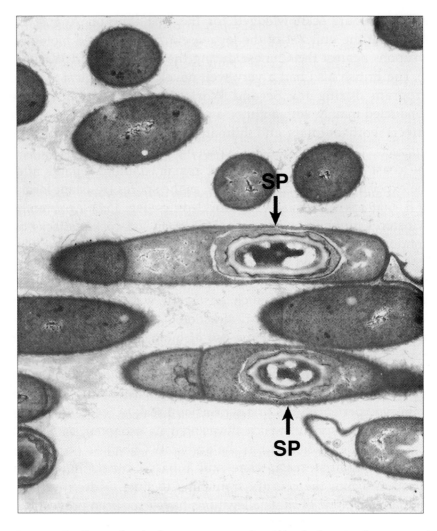

Fig. 15.1. *Bacillus anthracis* showing a spore (arrow) which is in the centre of the bacterium. ×14000. TEM. SP = spore.

During the 1980s, the US administration alleged that the former USSR was continuing with research in biological weapons although it is signatory of the 1972 biological weapons convention. A major purported violation was an explosion of a biological weapons facility in Sverdlovsk in 1979, causing an outbreak of anthrax in the region. Although USSR denied it, a study released in 1994 by the US and Russian scientists concluded that the anthrax outbreak was caused by a leak from the Sverdlovsk factory.

The apartheid regime of South Africa was the first to test the transmission of *B. anthracis* enclosed in letters. This disclosure was recently made during the trial of Dr. Wouter Basson, the then chemical and biological warfare expert of the South African government. In addition, he also tested *B. anthracis* to see whether they could survive temperatures of more than 180°C in a burning cigarette. The spores of this bacterium did survive, and Dr. Basson and his team toyed with the idea of sending gift packets of these cigarettes to anti-apartheid exiles. However, there is no record to prove that this had actually happened.

In 1950, the Australian government introduced myxomatosis virus in an effort to control the rabbit population in that country. Within one year almost 99% of the rabbits in Australia died. But gradually the survivors who had become immune to the virus multiplied and the problem re-emerged. Australian scientists have recently created a deadly mouse virus by genetically altering the mousepox virus. It is so virulent that even the mice immunized against the natural mousepox die. This shows how dangerous genetically engineered microbes can be, and it is quite possible that genetically engineered microbes for use against humans already exist in some countries or could be created in the future.

Biological warfare has an enormous destructive potential, especially against civilian populations and is a glaring example of man's inhumanity to man. On the one hand, the scientists are struggling to protect humans from the ravages of infections, and on the other hand, research continues to weaponize and produce more deadly microbes. This paradoxical behavior is a result of hatred implanted in our minds against people of other races, religions and ideologies. Unless hate is replaced with tolerance and understanding, humans will continue to use biological and other forms of weapons, irrespective of paper treaties and prohibitions that exist at present.

Anthrax

The Bacterium

Bacillus anthracis is a gram-positive aerobic, spore-forming bacterium found in the soil. In the animal body, the bacillus occurs individually or in pairs, but in cultures they form long curly chains, giving the colony a "medusa-head" appearance. The spore is the infective stage of the parasite and it can survive in the soil for an indefinite period. Herbivores are the most susceptible species and acquire anthrax when grazing on contaminated ground. Normally, humans acquire infection directly from infected animals or indirectly from contaminated animal products, such as wool.

The Disease

Natural infection from anthrax can occur from contaminated animal products. The disease can manifest itself in three ways, involving the skin (cutaneous anthrax); minor skin abrasions produces skin infection, which begins as a pimple and grows rapidly in two or three days into an ulcer. A dry black scab usually forms with edema or swelling around it. This condition is known as hide-porter's disease. Inhalation of anthrax spores can cause serious pneumonia, also known as wool-sorter's disease, which may result in death within 2–3 days. Intestinal anthrax is rare and occurs on ingestion of infected meat and also carries a high mortality rate. Occasionally, anthrax involves the brain and this too can rapidly lead to death.

＊＊＊＊＊

"Men marched asleep. Many had lost boots,
But limped on, blood-shod. All went lame, all blind;
Drunk with fatigue; deaf even to the hoots
of gas-shells dropping softly behind.

My friend, you would not tell with such high zest
To children ardent for some desperate glory,
The old Lie: Dulce et decorum est
Pro patria mori".

Wilfred Owen

(The last lines in Latin mean 'It is a sweet and seemly thing to die for one's country'. Wilfred Owen describes the First World War, in this poem, in which he himself died.)

16

Sex and Survival

The male *Latrodectus mactans* moves cautiously, very cautiously, towards his female counterpart. As he comes nearer, his nervousness increases, as making love to her may mean death. In contrast, the female of the species sits patiently in her place, eyeing every move that he makes. Plucking up courage, he suddenly jumps on her and copulates in the special way that spiders do. Once the act is over he runs for his life, as he is defenseless, unlike his female counterpart, who is armed with a fang. But alas, the bigger and stronger female easily grabs him, and pierces his body with her fang. The injected poison, stronger than that of a rattlesnake, instantly kills him. She then devours him, piece by piece, providing much needed nutrition for her unborn progeny. This macabre behavior is typical of the Black Widow spider, which is found throughout the warmer regions of the world and is easily identifiable by her glossy black color and an hour-glass shaped red mark on her abdomen.

Reproduction is the ultimate objective of all life on earth. When an animal reproduces, its genetic code or DNA is passed on to its offspring, so that the species survives. At an individual level, once the task is over, the male becomes redundant; he either dies or is disposed off. This seemingly cruel behavior is particularly common in insects and is not confined only to the Black Widow spider. For example, the Praying Mantis (*Mantis religiosa*) also behaves in a similar manner. The female, like the Black Widow spider, is also much larger and stronger than the male. In an attack position she folds her two front legs towards her chest, giving the appearance of "praying", while she is actually positioning herself to attack another insect. The Praying Mantis is cannibalistic, and often eats her male partner, beginning with his head, even as the male's hindquarters are still engaged in copulation. The loss of head does not interfere in the males' copulatory activity.

The Honey Bees, whose behavior has been studied a great deal by the scientists, are social insects. Their colonies are made up of a queen, several thousand workers and a few drones (male bees). Both the queen and the workers are females, but only the queen is able to mate and lay eggs. The population sex ratio is highly biased in favor of the females, as only a few males are needed for mating with the queen. The population of the drones is kept low in the colony because they develop from a few unfertilized eggs, produced by the workers, as compared to the larger number of fertilized eggs produced by the queen. The drones which succeed in mating with the queen die, as their reproductive organs are torn during the process. The unmated surviving drones are driven out of the hive by the workers, and die in the wilderness without food or shelter.

The fire-flies, which shine like stars in the night, can play an amazing game of deception with the reproductive process. Each species of fire-fly has its own light flashing pattern, to attract a member of the opposite sex for mating. However, a species of fire-fly (genus *Photuris*), has developed the ability to mimic the flashing pattern of another species. The female uses this mimicry to attract the male of the mimicked species. When these males approach her for mating they are pounced upon and eaten.

The deep-sea Angler-fish has a most unusual mating behavior, which probably developed as a result of the darkness that prevails at those levels. The male, once it locates a female, immediately attaches itself to her body by his mouth. The site of attachment is non-specific. After some time the lips and tongue of the male fuse with the flesh of the female. Gradually, all the tissues of the male degenerate with the exception of its reproductive organs. The vascular system of the two fishes then coalesce and the blood of the female nourishes the male. The male thus becomes an appendage of the female, without a separate existence, and its function becomes restricted only to the supply of sperms for fertilizing her eggs.

The Starfish is one of the most well known creatures of the sea, as it is very widespread in the oceans of the world. It is so-called, because of its five or six arms, giving it a star-like appearance. A French scientist first observed that in a species of Starfish (*Ophiodaphne materna*), the female actually carries the smaller male in her mouth, so that his sperms are available whenever she wants her eggs to be fertilized. A portable male in the service of the female!

Another example of female dominance, is seen in the Spoonworm (so-called because of its shape), belonging to the genus *Bonellia*, which is also an inhabitant of the sea. The female of this species can be as large as 2 meters, while the fully-grown male does not exceed 2 mm, and lives inside the uterus of the female! This unique site of location begins when a *Bonellia* larva is picked up by the proboscis (the long bifurcated end), of the female worm from the sea bed. The larva then migrates to the uterus of the worm where it grows into an adult male. It has a rudimentary mouth, which is no longer used for eating, but for discharging sperms. It now becomes completely dependant on the female for its survival, and is nurtured by her for the sole purpose of fertilizing her ova.

In some animals, the need for fertilization by the sperms has been totally eliminated, and the unfertilized ova can develop into a new individual. This results in the formation of a clone or a replica of the mother, as there is no male input. This type of reproduction is known as parthenogenesis, and is seen in a number of insects, frogs, fishes, salamanders and lizards.

Parthenogenesis in the whiptail lizard (genus *Lacerta*), has been studied in some detail and shows an interesting mating pattern.

Although all the individuals in this species are females, an isolated female cannot reproduce on her own. An egg-laden female solicits courtship from another female of its own species. A female which is not ovulating (a period of one or two weeks each month), responds by performing pseudocopulation, i.e. producing the typical male ritual by mounting and wrapping around her. During this process, the egg-laden female discharges her eggs. A few weeks later their roles may be reversed. This observation indicates that sexual reproduction was a part of their previous existence and is still coded in their genome, but later disappeared due to the non-availability of the male, or for some other reason.

Another twist in the pro-female drama of nature is that there are microorganisms mostly belonging to the genus *Rickettsia*, which selectively kill the male, but not the female offsprings, of certain insects. *R. tsutsugamushi*, which causes scrub typhus in humans, has been implicated as a cause of the all female-progeny of its mite host. John Werren of the Department of Biology, University of Rochester, who has studied this phenomenon, states that infection by microbes distorts the sex ratio in many insects, but always in favor of the female. It appears that this kind of killing mechanism is developed to get rid of the males, which may become an unnecessary drain on the resources of a particular insect colony.

Recently, Orly Lacham-Kaplan of Melbourne's Monash University, has shown that female mice can conceive through the use of cells other than sperms. She and her group injected mouse oocytes (eggs), with cells from adult mouse tissues, to see if they would fertilize oocytes and support embryonic development. Fertilization occurred, thereby obviating the need for sperm to do this job. In theory, the technique could be extended to humans leading to "Sperm-Free" conception! A Japanese scientist Tomoshiro Kono and his colleagues have also produced parthenogenetic mouse. They call it "Kaguya" after a Japanese fairy tale character.

Golden Lance Head snake of Brazil has the best of both worlds. It has 3 sexes: male, female and hermaphrodite.[a] The presence of hermaphrodite ensures that the absence of the opposite sex does not prevent the propagation of species.

In trematodes all worms are hermaphrodites, with the exception of schistosomes. Schistosomes live inside the blood vessels of the host and the male and female remain in copulation all their life, which may be many years. Living in the vast maze of blood vessels, it makes sense to stick to your partner once you have found her!

The Aphids can give birth to fatherless offspring, each one a clone of its mother. The embryo inside her womb may have an even smaller embryo in her womb and so the Aphid gives birth to daughter and grand-daughter at the same time. As no male is involved, time is not wasted in mating and a large population size is quickly reached.

Does all this mean that the male is becoming redundant? Not really. Parthenogenesis is an economical way of reproduction, as it avoids mate searching and selection, but the resulting clones have a hazardous existence. For example, if a new disease appears, which is capable of killing one member, then all members may die due to their equal susceptibility. This is a recipe for eventual disaster. Observations reveal that all asexual species, with the exception of microorganisms, are relatively new because most have perished during the course of evolution for the above reason.

In contrast, sexual reproduction, in spite of its disadvantages, is very widespread in nature, as it ensures long term survival. This is because sex leads to the combination of two separate sets

---

[a]The word hermaphrodite comes from Greek mythology. Hermes, the god of dreams and guardian of livestock, fell in love with Aphrodite, the goddess of love. Her first son Priapus was profoundly ugly but her second son had matchless beauty. As he was bathing one day he was seen by a nymph, who fell in love with him. She begged to be allowed to unite with him, so that they are never ever separated. Her wish was granted by Zeus, and thus a male-female combination or a hermaphrodite was born.

of genes and this creates limitless variations. No single disease or environmental change (unless it is extreme), can wipe out the entire species. Evolution works to the advantage of the population with genetic variations. It is for this reason that sex will continue to play a vital role in the survival of life on earth and the male will always remain in demand. However, in terms of relative biological importance, the female is clearly the winner, although human males often think otherwise.

How Males and Females are Formed

Each cell in our body receives a complete set of hereditary instructions. These instructions are contained in DNA, which is packaged in a structure known as a chromosome. Each human cell has 23 pairs or a diploid (double) number of chromosomes, i.e. 46. The genetic make-up required for making a male is X and Y chromosome on the 23rd pair, and for making a female, two X chromosomes (Figs. 16.1 and 16.2). The X and Y chromosomes are carried by the sperms. An X-carrying sperm will produce a female (XX) embryo, and a Y carrying sperm, a male (XY) embryo. The female oocyte (egg), does not determine the sex of the embryo.

Most of the differences between females and males are a secondary consequence of the absence or presence of the Y chromosome. The Y chromosome determines that the testes, rather than the ovaries, form in the developing embryo. The testes once formed, produce the male sex hormone (testosterone), which gives the male configuration to the genitalia. If there is no male hormone, then a female will be formed.

It is possible to detect the sex of unborn fetus by amniocentesis, when amniotic fluid is drawn out from the ammotic cavity. The Y chromosome can be identified in the cells from the fluid.

Recently, Denis Lo and his colleagues from the Chinese University of Hong Kong described a non-invasive method of gender diagnosis of fetus. This is based on the detection of fetal nucleic acid in maternal plasma which is released from the placenta. In other words it should now be possible to determine the sex of the unborn fetus by examining mothers blood.

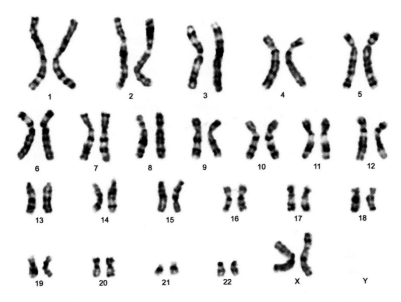

Fig. 16.1. Female chromosome pattern. Note the absence of Y chromosome. (Courtesy of AKU)

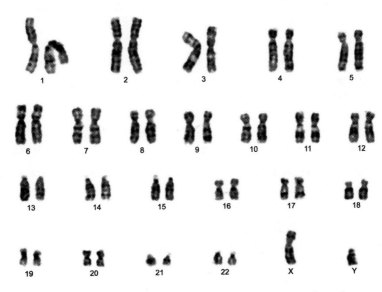

Fig. 16.2. Male chromosome pattern. Note the presence of Y chromosome. (Courtesy of AKU)

✳✳✳✳✳

"For even as love crowns you
so shall he crucify you.
Even as he is for your growth
So is he for your pruning".

Khalil Gibran

From "Love", Chapter 2 of *The Prophet*, 1923.

17

Do Microbes Possess Consciousness?

"What is the ultimate scope of science? Is it just the *material* attributes of our universe that are amenable to its methods, whereas our *mental* existence must forever lie outside its compass?" This question is raised by Roger Penrose in his book, *Shadows of the Mind*. The answer of course is that there is no limit to the scientific domain. Consciousness is probably the most important issue of our existence, which needs to be more extensively researched by scientists all over the world.

How do we define consciousness, when in reality we do not understand its origin or its nature? As far as we know, consciousness is subjective, and what is consciousness to us, may not be the same for a plant or an amoeba. Yet there appears to be some sort of consciousness, which is universally present. One of the criteria used by biologists for defining life is its capability to respond to stimuli. How can one respond, unless one is conscious? Is it necessary for an organism to have a brain like us to be conscious? If we think beyond the "brain-consciousness" paradigm, then the problem becomes easier to handle. We can then explore the natural world, from the higher animals to viruses, on an equal footing.

Let us consider some examples. The Rabies virus enters the body via the saliva, which is introduced by the infected animals (mostly dogs), into the bite wound. The virus travels along the nerves of the bitten animal to the brain. It then selectively concentrates in the limbic system, where the emotional centers are located. This causes a major change in the nature of the animal. A dog which was docile and obedient becomes aggressive and uncontrollable. It bites every animal in its vicinity including its owner. This is the only way the virus can survive and propagate itself. How did the virus trace this complicated pathway to achieve its aim?

A protozoan parasite, *Sarcocystis*, is very widespread in nature and involves the voluntary muscles of the body. A species of *Sarcocystis*, *S. singaporensis* has a rat-python cycle. The parasite develops in the python intestines, producing the infective stage (oocyst), which are passed in the python feces. Rats eat the python feces, because it contains a lot of organic matter, and in the process get infected. The parasite then begins to develop in the rat muscles and makes them partly dysfunctional. Some rats develop gangrene of the limbs and also become blind. In short, the rat usually survives, unless it has ingested a large dose of oocysts, but it cannot run as fast as it would normally. It becomes an easier prey for the python and the parasite is able to complete its life cycle.

Filarial worms are nematodes (round worms), found in many parts of the tropics. They are dependent on insects for transmission from one host to another. *Wuchereria bancrofti*, a common filarial worm of humans, is transmitted by the mosquito. During their life cycle, microfilaria (embryos) circulate in the blood, waiting to be picked up by the mosquito. As the mosquitoes transmitting this parasite are night biting, the microfilaria appear in the blood during the night. If they were to appear during the day, they would not get picked up by the mosquito. If they circulate both during the day and night, they would be wasting a great deal of energy. Similarly, there is another filarial worm called *Loa loa*, its microfilariae appearing during the day, because the vector is a day-time biting fly (*Chrysops*). Many biochemical explanations have been given to explain microfilarial periodicity. But it largely remains an enigma and an amazing demonstration of a parasites capability to regulate its behavior to the feeding habits of its vector.

There is a trematode (flat worm), called *Leucochloridium macrostomum*, which has snails as its intermediate host and song birds as its definitive host. The terrestrial snail is usually hidden in the vegetation and after getting infected by the parasite, there is a marked change in its behavior and appearance. The infective stage (*cercaria*), of the parasite migrate to the snails tentacles, which swell enormously, become bright green in color and begin to pulsate at 50–100 beats per minute. The brightly colored pulsating tentacles attract the song birds which eat up the snails. The song bird is the final destination of the parasite and by this extraordinary means it reaches its goal.

Polymorphus minutus is also a parasite of the birds and its intermediate host is a shrimp. The normal color of the shrimp is brown or yellow, but after getting infected by the parasite, it becomes blue. It has been shown experimentally, that birds are more attracted to blue rather than brown shrimps. The change in the intermediate host's color, therefore, increases the possibility that the definitive host (bird) will get infected. It almost seems as if *Polymorphous* is aware of this fact and brings about the change in color.

Fungi of the genera *Hohenbuehelia* and *Pleurotus* have devised traps to catch nematode larvae that may crawl over their fruiting bodies. They produce micro drops of an acid at the ends of their aerial hyphae and as soon as the nematodes come in contact with the acid, they get paralysed. The hyphae then enter the nematodes mouth and anus and eat up their tissues.

Like parasites, bacteria too show awareness of their surroundings. If offered sugar solution on the one hand and acid on the other, they would swim towards sugar and away from acid. Similarly, they can sense temperature, light and even magnetic fields. During conjugation (mating), they pass on to their susceptible brethren antibiotic resistance to enable them to survive exposure to antibiotics. Is this some sort of altruistic behavior?

About ten years ago, John Cairns and his group at the Harvard School of Public Health, reported that bacteria tend to show mutations likely to benefit them when put under stress by starvation. This is known as "directed mutation", which goes against the widely held belief that mutations are always random. Other workers have suggested that under stress, "hyper mutation", rather than "directed mutation" occurs. Even so, the bacteria are responding in a manner which gives them the best chance of survival.

Many antibiotics act when bacteria are dividing. Stanford University researchers recently showed that exposure to certain antibiotics triggers an SOS response in bacteria, resulting in the shutdown of DNA replication and transient dormancy, thus enabling their survival.

The scientists from the Weizmann Institute of Science in Israel reported that when the process of amoebic division gets stalled during separation, the amoeba gets help from their neighbors. An amoeba moves in to separate the dividing cells. The researchers noted that this is not an isolated case of help and this kind of "amoebic midwifery" is a routine happening.

To think that only we, the humans, have consciousness, has alienated us from nature and has led us to mercilessly exploit and destroy our eco-system. If we accept that consciousness is universal, then our relationship to the entire animal and plant world will change and we will give up the idea that man has dominance over all the inhabitants of the earth and that they were created to serve him.

✳✳✳✳✳

"Life, and all that lives,
is conceived in the mist,
and not in the crystal.
...Yet you shall not deplore having known blindness,
nor regret having been deaf.
For in that day you shall know the hidden
purposes in all things.
And you shall bless darkness as you would bless light".

Khalil Gibran

From "The Farewell", Chapter 28 of *The Prophet*, 1923.

18

Nature's Clocks

In 1729, a French astronomer by the name of Jean Jacques de Marion, was intrigued by a plant in his garden, which opened its leaves in the morning and closed them in the evening. He wondered if the widely held belief that this is triggered by the appearance and disappearance of the sunlight, was correct? To test this, he transferred the plant to a dark room. The plant continued to open and close its leaves at exactly the same time! This proved that there was a sensitive timing mechanism within the plant, which operated independently of the sunlight.

In 1737, a Swedish botanist, Carolus Linnaeus, made a similar observation. He noticed that the flowers of different plants opened at different times, and this process was so accurate that one could even tell the time of the day! It seemed as if the plants had internal clocks!

More than 200 years passed since these observations were made before scientists discovered that the internal clocks are indeed present, and found not only in plants, but also in all animals. This led to the birth of a whole new science of biology, known as chronobiology (chrono = time), which is essentially a study of body functions in relationship to time.

Chronobiologists first concentrated their attention on finding the location of the so-called internal clock or clocks. Many animals were used for this purpose, but the fruit fly (*Drosophila melanogaster*) was the most useful, as it was easy to breed in large numbers, and had a sleep-wake cycle like that in the higher animals, which is presumably controlled by an internal clock. It was discovered that animals have two types of internal clocks, a central clock in the brain, and peripheral clocks in other organs of the body. These clocks work in unison. The destruction of the central clock makes the peripheral clocks inoperative, but the destruction of the peripheral clocks did not impair the working of the central clock.

The genome of the fruit fly has recently been sequenced and the gene responsible for the programming of the clocks has been identified. The same "clock gene" is found in all mammals and humans, which means that it is a highly conserved gene, and important for the survival of all animal species.

We now know that the central clock in mammals and in humans is located in a group of cells in the hypothalamus just above the optic chiasma (the crossing of the optic nerves at the base of the brain), and called the suprachiasmatic nucleus (SCN). The light sensors found in the retina have a pathway to the SCN. The retinal cells convey the information about the ambient light to SCN, and SCN interprets it as day or night. SCN then sends the information via the spinal cord to the pineal gland (so named because it resembles a pine-cone). The pineal gland secretes a hormone called melatonin in response to the message (Fig. 18.1). If it is night, melatonin secretion rises and sleep is induced. If it is day, melantonin secretion decreases and the person awakes. In one of the most exciting discoveries in the field of chronobiology, the light sensors in the retina have recently been identified by David Berson, at Brown University. Dr. Berson said that a deeper understanding of the light sensors might lead to a novel treatment for disturbances of the body's internal clock. It may turn out that the people who have defective light sensors could suffer from "time blindness", similar to color blindness, causing sleep problems.

The sleep-awake cycle varies a great deal from animal to animal. For example, in giraffes the sleep time is estimated to be only 1.9 hours, while in bats it is about 19.9 hours. In humans it is 8 hours. During sleep, the body temperature, urine production

Section of Human Brain

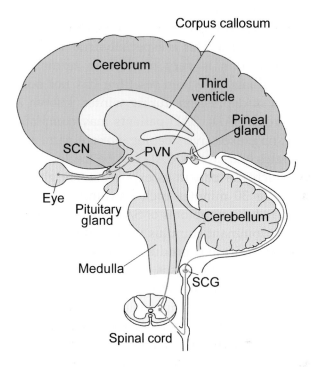

Fig. 18.1. The diagram shows the neural pathway between the eye and the pineal gland. The pineal gland secretes melatonin which induces sleep.

SCN = Suprachiasmatic nucleus   PVN = Paraventricular nucleus
SCG = Superior cervical ganglion

Adapted from *Evolving Brains* by John Allman. Publisher: Scientific American Library, New York. (Original drawing by Joyce Powzyk with Permission of Henry Holt & Co., LLC)

and the secretion of various hormones drop under the influence of internal clocks.

The jet lag that occurs during air travel is because it creates disharmony between SCN timing, and the light cue which it receives from the retina, the passenger being in a different time zone. It may take a few days for the SCN to get reset to the new timing, but before that happens, the sleep pattern becomes very irregular. Administration of melatonin has been shown to hasten the resetting of the SCN.

SCN also needs resetting during seasonal changes from summer to winter and vice versa. This is especially true in countries where the seasonal variation is very marked in terms of light and darkness. Individuals, in whom resetting has not occurred, may develop a form of depression during winter, known as Scasonal Affective Disorder (SAD). SAD patients may complain of lack of energy, overeating, excessive sleeping, weight gain, and a craving for carbohydrates and sweets. SAD can be treated by bright light therapy, in which the patient is exposed to bright light (10 000 lux intensity), for 30 minutes each morning. The light helps in resetting SCN and the symptoms of SAD abate or disappear. Light is therefore the dominant stimulus which affects the SCN, and as the earth rotates on its axis producing alternating day and night, a rhythmic activity results. This is known as the circadian rhythm, from *circa*, meaning about and *diem* meaning day, as the circadian rhythm controlled by SCN is approximately 24 hours.

In January 1998, scientists from the Laboratory of Human Chronobiology at Cornell University in New York, discovered that by shining a bright light for three hours on the back of the knees, they could reset the SCN by a few hours. This may turn out to be a useful technique for overcoming the jet lag and future airplanes may provide a light source placed under the passenger seats for this purpose!

It is common knowledge that some individuals are alert and productive at midnight, while others can't keep their eyes open at this time. This could be connected to our genes, as some findings have shown. Recently, a female patient came to the University of Utah, in Salt Lake City, USA complaining that she falls asleep at 7:30 pm and wakes up around 4:30 am and all the members of the family have the same problem. All of them were suffering from a condition known as familial advanced sleep phase syndrome (FASPS). Later studies on this family, done at the

Howard Hughes Medical Institute, revealed that all the family members had a mutation in one of their genes. This mutation set the SCN a few hours earlier than the normal.

The aging process adversely affects SCN functioning as some of the neurons in SCN degenerate. As a result melatonin production decreases. The elderly (over 65), are therefore more likely to suffer from sleep disorders. The problems include frequent daytime naps, less total nighttime sleep and a phase shift causing them to go to bed early and get up earlier, as in FASPS.

The internal clocks regulate the reproductive cycle in many animals. In a series of remarkable experiments, male hamsters were shown to change their testicular size, and thus their reproductive capability, in response to the length of the day. Hamsters breed and bear their young during spring and summer, when the days are long, so that the newly-born are able to move around and find food. As the days shorten, the male hamster testes shrink and spermatogenesis stop. The cessation of reproduction occurs at the time of the year when the survival of the young could be jeopardized by unfavorable weather conditions and difficulty in finding food. When Tennyson wrote, "In the spring a young man's fancy lightly turns to thoughts of love". It may not be just poetic imagination!

Research now reveals that the administration of medicine at the right time makes them more effective because disease manifestations are also controlled by internal clocks. Medications for asthma, cancer, epilepsy, cardiovascular disease, and allergies all have shown better results, if administered at the right time. The right time is the time when the drug is most needed; for example, the blood pressure is at its highest in the morning, which is why heart attacks are most likely to occur early in the day. To give a constant dose of medication will push the blood pressure lower than needed during the night, without sufficiently reducing it during early morning. Similarly, many of the asthma patients suffer most at night at about 4:00 a.m., so the appropriate time for medication would be 3:00 a.m., to prevent an attack, rather than at 8:00 p.m., which is usually the case. Rheumatoid arthritis hurts most in the morning hours, when the body's natural anti-inflammatory agents appear to be low. This would therefore be the best time to take aspirin or other pain killing drugs.

We have come a long way from de Marion's observations made on plants in 1729. We now know the location of most of the internal clocks in humans, and we also know that they have a

---

Other Rhythms

In addition to circadian rhythms which span 24 hours there are other types of rhythms which affect the lives of many organisms including humans. These are

1) *Ultradian rhythms*

These have a period shorter than 24 hours. Heartbeat and respiration rate are obvious examples. Adults normally have a heart rate of 60–90 bpm (beats per minute) while resting. The respiratory rate in adults is 10–15 breaths/minute.

2) *Infradian rhythms*

These have a period longer than 24 hours and may range from a few days to many years. Amongst the longest, of such cycles is that of the Cicadas, which are members of the order Homoptera and similar to Aphids. Same species of Cicadas emerge from the soil every 17 years and survive as adults for a day or so, during which time they mate, and die. The eggs laid by the female will re-emerge after another 17 years! How they maintain this extraordinary timing is not known.

There are Cycles which coincide with the waxing and waning of the moon and are known as *circalunal rhythms*. The human menstrual cycle is the best known example of this. The Lunar cycle also affect the tides and produce *circatidal rhythms* which affect many littoral animals. An example of a circatidal rhythm is the marine diatom *Hantzschia* which at high tide descends into the sand and at low tide rises to the surface.

In theory one could extend the time related cyclical pattern to the movement of the planets, stars and all the heavenly bodies.

profound effect on virtually all physiological activities. However, we still do not know how to reset the clocks. For example, if we are able to reset the clocks responsible for aging, we may not only be able to extend our life span but also overcome many age related diseases. There is no doubt that chronobiology has the potential of bringing great benefit to mankind and many new discoveries are waiting to be made in the future.

✳✳✳✳✳

A Time for Everything

To everything there is a season, and a time to every purpose under the heaven:

A time to be born, and a time to die;
A time to plant, and a time to pluck up that, which is planted;
A time to kill, and a time to heal;
A time to break down, and a time to build up;
A time to weep, and a time to laugh;
A time to mourn, and a time to dance;
A time to get, and a time to lose;
A time to keep, and a time to cast away;
A time to rend, and a time to sew;
A time to keep silence, and a time to speak;
A time to love, and a time to hate;
A time of war, and a time of peace.

Ecclesiastes: 3:1–8

19

The Plant Kingdom

Plants evolved from ancient green algae and probably appeared on earth about 400 million years ago, paving the way for the subsequent appearance of higher animals. The oldest flower fossil was discovered in northeast China recently, and is 125 million years old. The oldest living plant is the Bristle Cone Pine, found in the White-Inyo region of California, which has an estimated life span of 4767 years. It has been named "Methuselah", after Noah's grandfather who supposedly lived for 969 years!

Plants figure prominently in the culture and religions of various nations. In several Greek myths, humans were transformed into trees, thus interlinking the two life forms. The Oak tree was particularly sacred to Zeus and other Greek gods. Some trees became sacred because of their historical association with humans. It was under the Pipal tree, common in India, that Gautama Buddha attained enlightenment (566 BC). Buddhists claim that the siblings of this tree, taken at that time, now exist in many countries and are treated by them with great reverence.

Plants are very important for the sustenance of life on this planet. They are the primary food source of all the animals, including the carnivores, as the carnivores survive by eating the herbivores. Plants also maintain the delicate gaseous balance between CO_2 and O_2 in the earth's atmosphere, by utilizing CO_2 and producing O_2. O_2 is needed for the survival of all animals, and for protecting us from the sun's UV rays by forming the ozone (O_3) layer.

Plants have one disadvantage over the animals in that most, but not all, are immobile, being rooted to the ground. They therefore require animal help for the pollination and dispersal of seeds. However, immobility allows them to conserve energy for reproduction and growth. Plants draw energy directly from the sun, using "solar cells" known as chloroplasts, which contain a green pigment, chlorophyll. Chlorophyll utilises CO_2 and water to form carbohydrates. Chloroplasts are membrane bound structures within the plant cell, and were originally a species of bacteria known as *Cyanobacteria*, which developed a symbiotic relationship with the plants during the course of evolution. The evidence for this has come from sequencing the chloroplast of DNA, which is very similar to *Cynobacteria* and quite distinct from the plant DNA.

For pollination to occur, pollen (analogous to the animal sperm), needs to be deposited on the sticky surface of the elongated stigma (female organ), which protrudes from the center of the flower. The shape of the pollen differs from species to species, and shows amazing beauty and symmetry under the scanning electron microscope (Fig. 19.1). In wind pollinated plants, such as the grasses, the stigma is hair-like, so as to catch the flying pollen as it sweeps across. The pollen, on getting attached to stigma, germinates and produces a "pollen tube" that runs through the plant's pistal to reach the ovaries at the base of the flower. Fertilized ova then develop into seeds.

For the pollen to be transferred from one plant to another, they have developed many ingenious methods. To attract insects, plants often produce colorful and sweet smelling flowers. The insects, which come in contact with the flowers, are further rewarded by a sugary nectar. Even acts of deception are employed for this purpose. The dragon orchid resembles a female wasp and emits a chemical, which simulates the wasp's pheromone. This attracts the male wasp, which tries to mate with the orchid, and in the process gets covered with the pollen. The frustrated wasp moves away and repeats its attempt at copulation with other dragon orchids thereby pollinating more flowers.

Fig. 19.1. Pollen of African Violet flower. Pollen of various plants show specific ornamentation and their small size enables them to become easily air-borne. ×8000. SEM. (Courtesy of Ms. Tan Suat Hoon, EM unit, Faculty of Medicine, NUS)

Not all flowers produce a sweet smelling odor. The world's largest flower, the *Rafflesia arnoldi*, has petals which are 1.6 ft long and the flower may weigh up to 7 kg. It grows in the islands of Sumatra and Borneo, and also recently flowered at the Kew gardens in the UK, where it drew large crowds as people came to witness this rare event. This flower gives off a very strong stench, which resembles that of a dead animal. This smell attracts certain species of flies, which normally go for carrion. The flies do not find any putrifying flesh inside the flower, but during the process of exploration get coated with pollen, and then transfer it to other flowers of the same species.

Some flowering plants and cacti get pollinated by a species of night flying bats. The flowers in this case bloom during the night and are iridescent white in color so as to become visible to the bats. This species of bats have long tongues with a brush-like surface, which they use to scoop up the nectar and pollen from inside the flower, and then transfer it to other flowers.

In general, cross-fertilization is preferred to self-fertilization in nature, as it ensures the mixing of the genes, thus preventing hereditary diseases and genetic defects. The mystery of how flowers accomplish this was solved recently by a team of scientists from the Cornell University. Working on the reproductive process of Brassica plants they found that the plants stigma can recognize "self" from "non-self" pollen. If "non-self" pollen lands on the stigma the usual process of fertilization proceeds and seed formation occurs, but if the self related pollen lands on the stigma, the cells recognize it, and the pollen does not grow. This recognition is based on a highly specific lock-and-key interaction, between the cell receptors on the stigma (lock), and the ligends on the pollen (key). The reaction is similar to the antigen-antibody reaction seen in animals.

The plants also have fascinating techniques for seed dispersal. The seeds are essentially small packages with all the genetic material (DNA), and a small amount of nutrition to initiate the plants development. The seeds are so well protected that they can often survive for months or even years. The Indian Lotus

(*Nelumbo nucifera*) seeds, from a drained lakebed in Manchuria, hold the record for maximum longevity. These seeds were about 1000 years old (estimated by radio-carbon dating) when first identified, but were successfully germinated on being grown in fresh soil.

The fruit is the seed bearing part of the plant, and is produced to help in the dispersal of seeds. The pleasant flavor, color and odor is meant to attract animals, which on eating the flesh of the fruit, disperse or swallow the seeds which are passed out in their excreta. In this respect the tropical fruit, known as Durian (*Duirio zibethinus*) is especially worth mentioning. Unlike other fruits, it emits a very pungent smell, which is recognized by many animals, and they travel from a great distance to eat the fruit. The Durian smell comes mainly from its very thick and spinous outer coat, which is so firm that no animal can easily break it. Only when the fruit ripens, does it drop from the normally tall Durian tree, and in the process breaks open. This exposes the fleshy seeds, which are then picked up by various animals, and are dispersed across the forest.

Coconut is another seed-carrying fruit of amazing construction. It is large but sufficiently light to float on the sea, which disperses it along the coastline in the tropics. Under its thin outer shell is a network of loosely packed fibers or husk, which traps the air in its meshes enabling the fruit to float. Inside the husk is an extra-hard shell, containing a rich store of food in the form of a white jelly-like material and about half-a-pint of water. The coconut water is nutritious and isotonic, providing the germinating plant with enough fluid to start its life.

The seed-bearing fruits are often hard and unpleasant to taste when unripe, to prevent animals from feeding on them prematurely. The mango, which when ripe is delectably sweet, is bitter and hard to eat when unripe and secretes a white irritating sap from its stem if broken.

The seeds of many plants do not spend their energy in producing fruits and simply get attached to the coat of moving animals and human clothing by their rough surface or hooks, and thus get mechanically transferred from place to place.

A few plants are self-reliant as far as dispersal of seeds is concerned. An excellent example of this is the Mediterranean Squirting Cucumber, which gradually fills its seed bearing pods with a slimy juice. Eventually, the pods rupture with a great force squirting the seeds as far as twenty feet away from the plant.

As stated earlier, plants form the basis of the food chain on which all life depends, but in exceptional situations, this relationship gets reversed and plants become animal eaters! Carnivorous plants are found in unfavorable environments with poor soil, such as bogs and marshlands. The "Marsh Pitcher" (*Sarcenia* spp), is a tube-like carnivorous plant occurring in Venezuela and other South American countries. It has a very simple mechanism to trap insects. At its top is a red hood, containing nectar-producing cells. The red color and smell of nectar attracts insects. In search of more nectar, the insects move down, and as a result slip into a pool of water at the base of the tube. Inside the water, they die and decay, enabling the plant to absorb the nutrients released from their bodies.

Many pitcher plants have slippery waxy inner walls and the insects simply slip into the bottom of the pitcher, being unable to hold on to the walls. It was recently discovered that a species of pitcher plant *Nepenthes bicalcarata* does not have a waxy interior and it allows the "scout" or investigating ants to get in safely. In turn, they bring many nest mates. Whenever it rains (which happens frequently in the tropics), the rim of the plant becomes a slippery slope and numerous ants slip in, as they had zeroed in thinking it would be a safe source of nectar.

Plants also have a few active insect killers and the most spectacular amongst these is the "Venus' fly trap" (*Dionaea muscipula*). It has kidney-shaped reddish leaves with margins fringed by a row of spikes. The nectar producing cells lie at the margin of the leaves to attract the insects. In the central part of the "trap" are bristles which act as a trigger to close the trap, when any insect touches them. Once this happens, the spikes firmly lock into each other and the insect cannot escape. After trapping the insect, the plant releases digestive enzymes, which disintegrate the insect tissues, and the nutrients are absorbed by the plant.

As in the case of the "Venus' fly trap", the sense of touch is highly developed in some plants. A common plant, which goes by the name of "touch-me-not" (*Mimosa pudica*) displays this quality. Its leaves are made up of a cluster of four leaflets, which when open are attractive to leaf-eating insects. However, on the slightest stimulation they fold up and take the appearance of a tangled weed. The baffled insect thus flies away!

The plants' defense against predators is more sophisticated than until recently realized. Ian Baldwin and his colleagues from the Max Planck Institute, Germany, found that plants can silently "communicate" with each other, and even summon help from caterpillar destroying wasps, when they come under attack by caterpillars. This they apparently do by emitting volatile chemicals, which carry the desired message. Similarly, some plants can raise their level of bitter chemicals or alkaloids in their leaves when attacked by predators. Astonishingly, it has been found that the neighboring plants also do the same, even before being attacked, by some mysterious inter-plant communication, which warns them of the presence of an approaching enemy.

Professor Tony Trewavas of Edinburgh University thinks that plants have "intelligence". He gives the example of a parasitic plant known as Dodder (*Cuscuta* sp) which attaches itself by coils to a host plant and within an hour detaches itself if the plant is not sufficiently healthy. It then tries another neighboring plant and continues to do so until it finds a plant to its liking. Trewavas thinks this is not a reflex action but a highly selective behavior, akin to an animal looking for greener pastures.

PLANTS AND MEDICINE

Plants have been the main source of medicines for thousands of years. Even as early as 50 years ago, the bulk of drugs listed in the International Pharmacoepia were plant derived. We still use plant-based medicines widely, but often in a more purified form. Chemists have produced many new drugs by modifying the original molecular structure to make them less toxic and more useful for humans.

Sadly, many valuable and potential sources of medicines are being destroyed by the cutting down of forests in the tropics. This is an irreparable loss, as some of these plants may become extinct and lost for ever. Following is the list of some important plant derived medicines:

Active Principles Derived from Natural Sources, that are Clinically Used as Drugs

Drug Name	Pharmacological Class	Plant Name
Artemisinin	Anti-malarial	*Artemisia annua*
Atropine	Anti-cholinergic	*Atropa belladonna*
Bromelain	Anti-inflammatory	*Ananas comosus*
Caffeine	CNS stimulant	*Coffea arabica*
Cocaine	Topical Anaesthetic	*Erythroxylon coca*
Codeine	Anti-tussive	*Papaver somniferum*
Colchicine	Anti-gout	*Colchicum autumnale*
Digoxin	Cardiotonic	*Digitalis lanata*
Ephedrine	Anti-asthmatic	*Ephedra sinica*
Morphine	Analgesic	*Papaver somniferum*
Nicotine	Ganglion stimulant	*Corydalis cava*
Papain	Enzyme	*Carica papaya*
Physostigmine	Cholinergic	*Physostigma venenosum*
Pilocarpine	Cholinergic	*Pilocarpus jaborandi*
Quinidine	Anti-arrhythemic	*Cinchona officinalis**
Quinine	Anti-malarial	*Cinchona officinalis*
Reserpine	Anti-hypertensive	*Rauwolfia serpentina*
Santonin	Anthelmintic	*Artemisia maritima*
Taxol	Anti-cancer	*Taxus breviofolia*
Theobromine	Vasodilator	*Theobroma cacao*
Theophylline	Anti-asthmatic	*Camellia sinensis*
Tubocurarine	Muscle relaxant	*Chondodendron tomentosum*
Vinblastine	Anti-cancer	*Vinca rosea*
Vincristine	Anti-cancer	*Catharanthus roseus*
Yohimbine	Psychogenic impotence	*Pausinystalia yohimbe*

*The spelling should have been Chinchona as it was named after the fourth Condesa de Chincho'n of Peru but Linnaeus in his classification left out the first "h" by mistake. (Courtesy of Dr. Anwar Gilani, AKU)

GM PLANTS

Genetically modified plants are being produced in many laboratories all over the world. The purpose of modifying the plant genes are manifold, which include high yield, disease and pest resistance, higher protein and vitamins content, herbicide resistance, drought resistance and frost resistance.

There is a great deal of opposition completely switching over to GM plants, due to a lack of sufficient knowledge, especially long term effects of the GM plants. As genes do not function in isolation, interaction with other plant genes may give rise to unexpected results. The GM plants may evoke allergic reaction and this has happened in case of transgenic soya beans containing Brazil nut gene.

The insect resistant plants were produced by inserting genes derived from a bacterium known as *Bacillus thuringiensis*, which produces a toxin that kills insects. The problem with this approach is that insects which are beneficial to the ecosystem are also killed. This in turn deprives birds of their food and thus has a wider implication on the eco-system, than previously thought.

The agro-chemical companies specializing in GM plants would obviously desire to profit as much as possible. To do this, they plan to introduce what is known as "terminator technology" — plants which produce sterile seeds, so that farmers have no choice but to buy seeds yearly from the producer companies. This has serious economic implications for the developing countries who want to use the GM crops.

In short, GM plants are potentially very useful but close monitoring will be needed before arriving at an international agreement on their wider usage.

✳✳✳✳✳

"I think that I shall never see
A poem lovely as a tree

A tree whose hungry mouth is prest
Against the earth's sweet flowing breast;

A tree that looks at God all day,
And lifts her leafy arms to pray;

Poems are made by fools like me,
But only God can make a tree".

Joyce Kilmer

From *Trees and Other Poems*, 1914.

20

The Long March of Evolution

The most profound study of evolution was first made by an English naturalist, Charles Darwin (1809–1882), whose recognition of relationship through shared descent ranks as one of the great ideas in the history of science, now proven by modern genetics. Animals may look different from each other, but from the fruit fly to humans, they use the same gene to establish their development pattern. This is called the *Hox* gene because it contains a region of conserved nucleotide sequence known as *homeobox*. The conserved nature of the homeobox sequence shows the basic commonality of all life forms.

The evolution of the visual system is another good example of interlinkage between all life forms. The metazoans have generated eyes of diverse nature ranging from simple light-sensitive photoreceptors to compound eyes of insects. Despite this great variation, all of them are generated by a single master gene *Pax-6*, which was first identified in mouse. Furthermore it has been shown that the induction of *Pax-6* in ectopic sites of *Drosophila* (fruit fly) can generate eyes on various parts of its body. Experimentally it is possible to produce *Drosophila* with eyes on its legs, wings and antennae!

Darwin grew up in a family of biologists and Charles's grandfather, Erasmus Darwin (1731–1802), was a well known doctor and botanist. During his student days at Cambridge, Darwin spent a lot of time collecting beetles and plants. He was already an experienced naturalist when he went on his famous journey on HMS *Beagle,* which sailed around the world from 1832 to 1836. While living aboard the *Beagle* he made copious notes of the fauna and flora that he observed. These observations led him to formulate a theory of evolution, which he put in a book form. The book was entitled *The Origin of Species* and was published in 1859.

His theory centered around three postulates:

1. Individuals in every population group show variants.
2. Some of these variations are inherited by their offspring.
3. In each generation, more offspring are produced than can survive. The survivors survive, because they possess favorable variants. This is the process of "natural selection" (Fig. 20.1).

Darwin produced direct evidence of his theory by showing how the finches of Galapagos Islands differ in their beak shape and size under selection pressure, to obtain various types of food, ranging from seeds to insects.

The English moth, *Biston betularia,* was another obvious example. The dark colored moths constituted less than 2% of the population prior to the industrial revolution. Following the industrial revolution, darkening of the buildings occurred due to soot, and the population of dark moths grew to 95%. The white moths were easily picked up by the birds against the dark background and were therefore almost all eliminated.

In modern times, one of the best examples of natural selection is seen in the common bacterium, *Staphylococcus aureus,* which inhabits our skin and sometimes produces serious disease. This bacterium was universally susceptible to Penicillin when this antibiotic was first introduced in the forties. Gradually, resistant strains of *S. aureus* started appearing and now, 60 years later, it is universally resistant. Because of the wide usage of Penicillin, natural selection eliminated all the susceptible strains. The end result is that *S. aureus,* as a species, survived because it had resistant variants in its original population, which multiplied under selection pressure. Evolution is, therefore, a preserver of some, and a destroyer of others.

Evolution tree

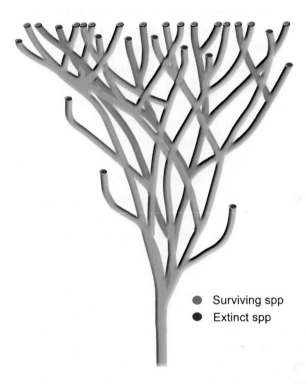

Fig. 20.1. Evolution proceeds more like branches of a tree than a ladder. During the process, many species become extinct due to a variety of reasons. Occasionally, mass extinctions occur, as it happened about 65 million years ago with the Dinosaurs.

Additional evidence in favor of change brought about by natural selection can be summarized as follows:

1. The fossil records of both animals and humans show distinct and progressive change, although complete phylogenetic trees are difficult to obtain for each species because of the time involved in the evolutionary process.

2. That changes have occurred is evidenced by the vestigial structures that exist in all animals. Humans have the tail bone (coccyx), a remnant of the ancestral tail, non-functional ear muscles, which were probably used for turning the ear towards sound, and a non-functional appendix. The appendix is a very important part of the intestinal tract in herbivores, where the cellulose-splitting bacteria reside. As humans became omnivorous the need for a large appendix disappeared and the organ became vestigial. Erector pili muscles in human skin, which produce goose pimples when we get excited, are similar to the erectile fur of other primates which is used for demonstration of alarm or aggressive intent. The goose pimples in humans no longer serve this purpose. In animals, there are many examples of vestigial organs. Some cave dwelling fish have eye sockets but no eyes, and flightless birds have useless rudimentary wings. Snakes have tiny hip bones and rear legs in the form of claws, as they have evolved from reptiles. Recently, whale skeletons have been discovered in Pakistan with legs, indicating that whales were at some stage terrestrial animals. The University of Michigan paleontologist who made this discovery has named the creature *Pakicetus*.

3. Selective breeding can bring about major changes in plants and animals. For example, a horse breeder can create faster horses by inbreeding the fastest amongst his stock. This happens because organisms differ from each other and the selected variations are inherited. In nature too, selective breeding occurs, which Darwin called *natural* selection, eliminating the weak and favoring the "fittest", the fittest generally being those who succeed in mating and passing their genes to their offspring.

There is unanimity of opinion amongst biologists that natural selection does occur, but can it *alone* account for the production of all the characteristics that we see in an array of animal and plant species? For this, we turn to Darwin's predecessor Jean-Baptiste Lamarck (1744–1829), for possible answers.

Lamarck was a firm supporter of evolution, and in his praise, Charles Darwin wrote in 1861:

> "...He (Lamarck) first did the eminent service of arousing attention to the probability of all changes in the organic, as well as in the inorganic world, being the result of law, and not of miraculous interpretation."

Lamarck is best known for his publication, *Theory of Inheritance of Acquired Characteristics*, which was published in 1801. He proposed that the changes brought about by the environment are passed on to the offspring, and it is the environment which brings about evolutionary changes. His theory centers around the principles of use and disuse. In our lifetime, if one regularly exercises, certain muscles develop. A black smith's arms are usually more well developed than his legs. Similarly, a runner's leg muscles are more developed than his arm muscles. Are these characteristics inherited? Are the offspring of a runner born with hypertrophied leg muscles? The answer is obviously no, but a Lamarckian could say that for such a change to occur it would take many thousands of years. There is no way to disprove this. The fact that Jewish children are born with a foreskin although circumcision has been practised for centuries does not disprove Lamarck because there is a difference between *external* damage and an *internal* stimulus to change.

There is an aspect of Lamarcism, which is quite intriguing. He spoke of animals striving towards a change or a "tendency to perfection". The giraffe eventually developed a long neck because it was trying to eat the leaves from tall trees. In a sense, the animal was "needing" to evolve in a particular direction, and there is therefore a mental component to the evolutionary change. One could say that if this was true, why hasn't the Eskimo developed fur like the polar bear? He hasn't developed fur, but certainly has a thicker layer of subcutaneous fat, as compared to the African, and has narrow eyes and double eye lids, to protect him from the glare of the snow. The Eskimo's needs have been fulfilled!

This could, however, have happened due to natural selection, as many evolutionary biologists would argue.

According to Lamarck, the "true principles" to be noted in his theory are:-

1. Alteration in the environment causes alteration in the *needs* of the animals.
2. Change in needs causes new activities to develop to satisfy those needs.
3. Every new need necessitates either more frequent use of some parts which develop and enlarge or the development of entirely new parts, to which the needs have imperceptibly given birth by effort of its "inner feeling". In short, the sequence is environment • need • new activities • physical change.
4. The opposite is also true. "The permanent disuse of an organ, arising from a change of habits, causes a gradual shrinkage and ultimately the disappearance and even extinction of that organ."

It is interesting that Aristotle, (384–322 BC) had suggested that there is a design in nature, but it is less from guidance from without, than the urges from within. It is the inner drive or "entelechy", which leads to the fulfillment or the fullest realization of a being.

There is a great deal of scientific evidence that the central nervous system has a profound effect on the immune system and the working of other organs of the body. For example, stress in animals can release cytokine (IL-I) from the macrophage, and these are the earliest molecules involved in the inflammatory response. The exam stress in students causes a significant drop in the number of natural killer (NK) and T-helper cells. Similarly, depression has an adverse effect on immunity. Mind/body researcher Lydia Temoshok studied the psychological factors associated with malignant melanoma. She found a correlation between emotions and the growth of cancer cells. We now realize that a vast number of diseases are psychosomatic in origin (psyche = mind, soma = body). In other words, thoughts and emotions do get translated into *physical* phenomenon.

Andrew Schwartz, a neurophysiologist at Arizona State University has recently reported that monkeys implanted with special electrodes in their motor cortex can move a cursor on a computer screen just by *thinking* about it. They were previously trained to use their hands for this purpose, but after their hands were strapped, they could still do it by using their mind.

If the mind and body are so intimately connected, why can't the mind also affect the gene, which is a physical component of the cell? This possibility cannot be totally dismissed. Darwin and Lamarck can both be correct.

Another giant in the field of evolution was Ernest Haeckel (1834–1919). Haeckel was trained as a physician but abandoned his practice in 1859, after reading Darwin's book *On the Origin of Species*, so as to concentrate on the study of evolution.

Haeckel agreed with Darwin's idea of evolution, but was not very enthusiastic about natural selection. Instead he believed that the environment had a crucial role in the formation of races, a view closer to Lamarck. He expounded the idea that "ontogeny recapitulates phylogeny". That is, embryonic development is a replay of that animal's evolutionary past (Fig. 20.2). We now know that this is not strictly correct, and new features unrelated to ancestry may appear or disappear during embryogenesis.

Nevertheless, there are structures, which clearly reveal evolutionary connections with the past. For example, all vertebrates start as a single cell, then become multicellular (morula stage) and then get curved with a protruding tail (tail-bud stage). In the tail-bud stage, all embryos have gills like that of the fish embryo. Later, the gills change into proper gills in fish, and gill bars in humans and other animals. The gill bars form the various structures of the face, mouth and ears. The embryos also develop fin-like structures, which turn into proper fins in the fish, and limb buds in other animals.

After birth, the human babies show features similar to those of other primates. Both are born with the grasp reflex, which

Embryonic devlopment of animals
by Ernest Haeckel (1874)

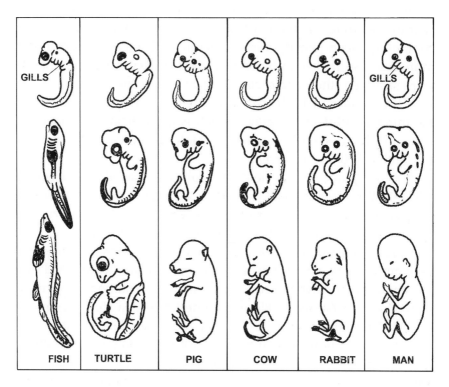

Fig. 20.2. We now know that Haekel's drawings were not very accurate, but he showed various animals, during their early embryonic stages. On this he based his theory that ontogeny (development of the individual) repeats phylogeny (history of the development or the evolution of the species).

disappears in humans, but persists in other primates. The reflex in the human baby is elicited when the baby's palm is stroked, to which it responds by making a fist. The reaction is so strong that the baby can be lifted by one's finger, if it is slipped into his or her palm. The grasp reflex in human babies disappears in about two months; during the course of evolution, it has lost its significance. But in other primates it is a life saving reflex. The baby monkey would fall to its death, unless it holds on to its mother's coat firmly while she is jumping from tree to tree. Therefore, the grasp reflex in the monkey persists, until the baby is strong enough to manage on its own.

According to the anatomist Tryphena Humphrey, of the University of Pittsburg, the first cutaneous reflex of the foot appears in the 10.5–11 week-old fetus, in the form of planter flexion of the toes. This reflex largely disappears in later fetal life, but is sometimes seen in newborn infants when it is referred to as a foot grasping reflex. This "undoubtedly represents an equivalent of the foot-grasp reflex found phylogenetically in birds and some arboreal mammals".

Human babies and chimpanzee babies are remarkably similar in their physiognomy. A newborn chimpanzee has big doleful eyes and a delicate face, but it later develops a protuberant face and massive jaws. The meeting point, as far as facial features are concerned, occurs in the first few months after birth, subsequent to which the species move in different directions.

Adult monkeys are quadruped; apes are partially biped; and humans are completely biped. But the human infant starts as a quadruped (crawls) and then gradually becomes a biped, again repeating its evolutionary past. Even in adulthood, humans involuntarily swing their arms while walking, in unison with their legs, as their quadrepedal memory has not totally disappeared.

An interesting evidence for Haekel's idea of evolution came recently from observations made by Gaeth, Short and Renfree of the Department of Zoology and perinatal medicine, University of Melbourne, Australia, and reported in the Proceedings of the National Academy of Sciences in 1999. It is generally believed that elephants were originally aquatic animals, and then they became terrestrial. In Melbourne University, they studied the histology of elephant embryos from animals culled at the Kruger National Park of South Africa. The embryos ranged from 0.04–18.5 grams, with estimated gestational ages from 58–166 days (full gestation in elephants is 660 days). The study revealed that the embryonic kidney of the elephants has a structure known as

Man's Antecedents

Primates are a diverse group of mammals which have close-set, forward-looking eyes and big toes and thumbs used for grasping. The digits have flattened nails rather than claws, as is seen in most mammals. Their brain is larger than any animal of comparative size.

Hominoids are a branch of primates, which separated into Pongids, Hylobatids and Australopithecines 25 millions years ago. Australopithecines were the intermediate species between the apes and humans, but unlike apes they were bipedal. Their brain size was 400–500 ml.

A scheme outlining possible evolutionary relationships between *Homo sapiens* and other living and fossil hominoids

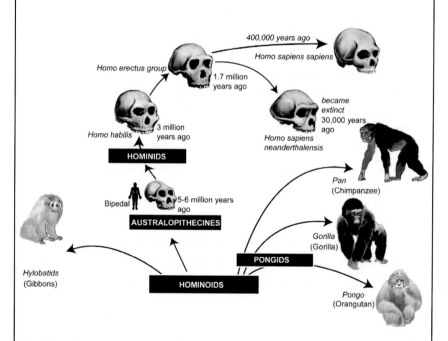

The Hominoids split into Pongids, Hylobatids and Australopithecines about 25 million years ago. The Australopithecines were the forerunners of *Homo sapiens* and were the first to become bipedal. (Diagram adapted from Gray's Anatomy, 38th Edition. Publisher: Churchill Livingstone. With permission from Elsevier Science).

Homo habilis, or the handy man, followed Australopithicines. He is regarded as the first true member of the human family. His brain had grown to 750 ml. The first recognized stone tools were made by him and he was a meat eater.

Homo erectus, or the upright man, followed *H. habilis*. His brain had grown to 900 ml, the speaking abilities had emerged, and they were able to manufacture multipurpose tools, such as axes. It is thought that *H. erectus* were the first to migrate out of Africa into various parts of the world.

Homo sapiens arose from *H. erectus* about 400 000 to 500 000 years ago, after passing through an archaic stage, to become *Homo sapiens sapiens*. The appearance of *H. sapiens sapiens* took about 150 000 to 200 000 year from the archaic stage. The brain size of *H. sapiens sapiens* grew to 1350 ml. The oldest fossil of the anatomically modern human is about 195 000 years, recently excavated from Ethiopia.

It is generally accepted that Africa is the birth place of the genus *Homo*, because of the high concentration of Hominid fossils found on that Continent. The mitochondrial DNA studies also indicate the African lineage of all the human races.

nephrostome, which is a tiny funnel-shaped duct, characteristic of aquatic animals. This structure was extensively present in the embryonic kidney for a few weeks, a sign that elephants had an aquatic past, which was being repeated in their terrestrial present. The authors also observed that the trunk appeared at the same time as nephrostome. It is therefore conceivable that the trunk evolved as an adaptation to the aquatic environment and was used as a snorkel when they swim in deep water, even till today.

Irrespective of the theories that exist about evolution, it is almost certain that no animal or plant appeared on earth in its present form. The long march of evolution, which started at the time when life began on this planet, goes on unabated. This is a fundamental law of nature.

The development of the brain was the key to the evolution of humans from hominids to *Homo sapiens*. Why and how did this extra ordinary development of the brain occur? If we bring back Lamarck into the picture, hominids used their brain more than any other animal and this in turn increased their mental capacity in later generations. However, there is a physical limitation on the growth of the brain imposed by the cranium (skull bone). To overcome this problem the brain has undergone infolding resulting in its present appearance. There is a limitation to this infolding process as well, but evolution will solve this problem too, as it has solved many other problems since we came into existence. Unless we indulge in an orgy of self-destruction, the future of *Homo sapiens* seems very bright.

✳✳✳✳✳

"Are God and Nature then at strife,
That Nature lends such evil dreams?
So careful of the type she seems,
So careless of the single life...

So careful of the type? but no.
From scarped cliff and quarried stone
She cries, "A thousand types are gone,
I care for nothing, all shall go".

Alfred Tennyson

(This is part of a poem *In Memoriam* (1850) written in memory of his friend Henry Hallam and includes the famous phrase, "Nature, red in tooth and claw")

21

Making Clones

The word clone stands for an exact copy or a replica of the original and is derived from the Greek word *Klon* which means a twig. The Greeks knew that a broken twig, when planted, grows into a species identical to its parent.

In the cloning process, there is either a partial or a complete transfer of nuclear DNA from one cell to another. In partial transfers, also known as molecular cloning, a short stretch of DNA is inserted into the host-cell, which is usually a bacterium or a yeast. The host-cell then expresses the substance coded by the inserted DNA. Using this technique, many useful chemicals and hormones have been produced *in vitro*.

Molecular cloning has been extended to produce transgenic (mixture of new genes with the animals' own genes) animals. For example, transgenic pigs have been created carrying human genes, so as to make their organs compatible with human tissues for transplantation. Similarly, cloned cattle have been produced to provide human antibodies for use against infectious diseases. Recently, Japanese scientists have developed transgenic silkworms that can spin cocoons containing human collagen. They have also produced a "blue rose" by mixing the genes of blue pansy with that of the rose! Numerous other uses for transgenic animals, particularly mice, are under investigation.

The shuffling of genes in this manner can also be done between plants and animals and vice-versa. In 1986, Japanese scientists were able to transfer gene coding for the substance luciferase, which causes glowing in the firefly, to a tobacco plant and the plant glowed in darkness! Conversely, plant-based vaccines are under investigation which would enable a person to develop immunity simply by eating a genetically engineered plant.

In cloning based on nuclear transfer, the whole genome is transferred, resulting in the production of a replica of the donor animal. This was first achieved in a mammal in February 1997, by Ian Wilmut and Keith Campbell from the Roslin Institute, Scotland, when they successfully cloned a sheep, named Dolly (Fig. 21.1). The question that Dolly was a genuine clone was settled by rigorous experimentation by mid-1998. Dolly produced a lamb, named Bonnie, proving that she was not physically or hormonally defective. However, fears were expressed that she might be ageing too fast as her genes originated from a six-year-old ewe.

Fig. 21.1. Sheep used for cloning Dolly.

How Dolly was Created

Three different sheep were utilized. These were the cell nucleus donor, the egg donor, and the surrogate mother.

A. **Cell Donor (Sheep I)**
Cells were taken from the mammary gland (udder) of a six-year old ewe and kept for six days in a medium with a minimal amount of nutrients. This was done to slow down the normally active DNA. The nucleus from the cell was then removed by micro-manipulation and the rest of the cell discarded.

B. **Egg Donor (Sheep II)**
An unfertilized egg was taken from another ewe and its DNA-containing nucleus removed but the outer membrane and the yolk were left intact.

C. The nucleus of **Sheep I** and the empty egg of **Sheep II** were brought together in the laboratory and given bursts of electricity to fuse them and to initiate cell division.

D. **Surrogate Mother (Sheep III)**
The multi-celled embryo thus created was implanted in a surrogate ewe with a black face, so that it would be clear that the offspring was not the result of a surreptious mating.

E. **Dolly (Sheep IV)**
After the usual gestation period, a baby lamb was born of identical color to Sheep I. This was named Dolly. DNA analysis confirmed that it was a clone of Sheep I.

Dolly was unique in that she has no father but three mothers, each contributing to her birth in some way. This would be a relationship very difficult for any human to handle!

Following Dolly's birth, scientists from various parts of the world have cloned mice, pigs and cattle. There is no doubt that the list will grow and it was recently announced that the scientists at the Oregon Regional Primate Research Center have produced two rhesus monkeys, "Neti" a female and "Ditto" a male, both by using nuclear transfer. Australian scientists recently claimed cloning a cow using a technique known as serial nuclear transfer. Under this procedure, the cells used undergo two rounds of nuclear transfer in the cloning process, instead of the normal one. Experimenters claim that this made healthier embryos and live births more likely.

Governments all over the world have opposed human cloning mainly for ethical and moral reasons. There is also medical apprehension about human cloning, as the technique is still in its infancy and the possibility of genetic defects and fetal death is very high. For example, Ian Wilmut's cloning experiment started with 434 sheep ova. Of these, 157 did not fuse with the donor cells and were discarded. Out of the 277 cells that fused successfully only 29 survived for transfer to surrogate mothers. Ultrasound scanning revealed 21 live fetuses, but all of these were gradually lost, leaving only one which was born normally and named Dolly.

Dolly's death was announced on 18th February 2003. For six years she was a symbol of success of a revolutionary development in the field of science. Her death was attributed to a lung infection which could not be controlled. It is still not known if her lung infection was due to some immunological defect she might have developed. She also had unexplainable arthritis. Dolly's relatively pre-matured death underscores how irresponsible it is to try and clone human beings given the current state of our knowledge.

The problems with human cloning are manifold. Even if born normal, the cloned human is likely to suffer a great deal of mental anguish, due to media attention and the complex relationship problems that would arise between him and the person he has been cloned from. Leaders of many religions regard cloning as a violation of human dignity and spirituality.

The supporters of cloning (not necessarily human cloning) point out that cloning is widespread in nature. All asexual form of division leads to clones. For example, bacteria are clones of each other. Cloning also occurs naturally in plants, worms, shrimp, snails, armadillos and some lizards. In humans too, cloning occurs in the form of uni ovular twins and nobody regards having twins as objectionable.

The supporters of cloning also point out that complete replication of a human being *both* in the physical and mental sense is not possible as the newly born clone will be living in a different time period from the donor and therefore be subjected to different environmental influences. Nobody knows exactly how much influence the environment has on the development of personality, but it is probably no less significant than the genes. It is therefore highly unlikely that an Einstein or a Hitler could be created by cloning, even if their tissues were available. In short, despite his genetic identity the cloned person is not likely to be identical to its progenitor.

Moving away from the debate on human cloning, it is clearly a useful procedure for selective plant and animal breeding. Almost all the orchids cultivated in farms are clones and so are many other flowers. In the near future many cloned animals, such as elite cattle and horses, would become available. Cloning could also be useful in reviving extinct or near extinct species as DNA can be extracted from dead animals. However, the recent effort to clone the Wooly Mammoth and the Tasmanian Tiger did not succeed. Chinese scientists are trying to clone the Panda and this may be the only way to save this species from extinction. Cesare Galli and his colleagues at the Laboratory of Reproductive technology in Italy have successfully cloned horse. The cells were obtained from the skin of two horses, a male African thoroughbred and a female Haflinger. This shows that champion horses and racers can now be cloned. The cloning could also enable gelding champion (castrated horses) to contribute to future generations. Finally, cloning is a useful tool for conducting basic research; to show how cells differentiate, how they age and die and how they can be revived or rejuvenated.

James Watt, the inventor of the steam engine, said in 1765, "nature can be conquered, if we can find her weak side". The modern genetics has found the "weak side" of how species are formed. It is up to us to make use of this knowledge in a responsible way without violating the rights of individual human beings.

＊＊＊＊＊

"By law of nature then art bound to breed,
that thine may live, when thou thyself art dead,
And so spite of death thou dost survive
in that thy likeness still is left alive".

William Shakespeare
Venus and Adonis

22

Replacing Body Parts

In Hindu mythology, Ganesh, a very popular deity, has a human body and the head of an elephant. He is the son of Shiva and Parvathi. On one occasion his mother Parvathi needed someone to guard the entrance while she was having a bath. She asked Ganesh to keep vigil. Meanwhile, Shiva her husband returned home after a long time only to be stopped at the entrance by Ganesh. Not knowing that Ganesh was his son, Shiva became enraged and slashed off Ganesh's head. When Parvathi came out of her bath she was aghast at the scene and became very angry with Shiva. To make amends, Shiva asked his soldiers to bring him the head of any living creature they first encountered. It happened that the first thing they saw was an elephant. Shiva transplanted that head on to Ganesh's body and this was how the elephant-headed Ganesh was born! In scientific terms, this would be classified as Xenograft and would rapidly be rejected by Ganesh's body.

The first recorded animal-to-human transplant (Xenograft) was done by a French surgeon, Matheu Jabouley in 1905. Jabouley used pig and goat kidneys and anastomosed them to the blood vessels of the arms of two patients who were in chronic renal failure. The kidneys functioned for approximately one hour after which they had to be removed.

Alexis Carrel, Jabouley's student, carried on with the transplant work in the US. He found that grafts worked well in dogs transplanted with dog kidneys, but not if the kidneys of another animal were used. He rightly concluded that the failure was not due to faulty surgical procedures, but was due to some immunity related factor or factors.

In 1964, American surgeons tried transplanting baboon organs into patients. The most famous of these attempts involved a Californian child called Baby Fae, who lived for 20 days after receiving a baboon heart after which it was rejected by the hosts tissues.

Primates are genetically close to humans, so their organs are less likely to be rejected. However, they also carry viruses which can spread to humans; hence they are rarely used.

Amongst other animals, pig tissues appear to be closest to humans and pig heart valves have been used to replace damaged human valves. In addition, transgenic pigs, which carry bits of human genes, have been produced and may become very useful for this purpose.

At present, only human donors are used for organ transplant. For this to succeed, the donor tissues need to be compatible with the recipient's. To test for compatibility, HLA (Human Leucocyte Antigen) typing and cross matching of tissues is done. Mismatching results in progressive attrition of the graft, leading finally to its rejection.

The credit for understanding the science of transplantation goes to a British immunologist, Peter Medawar (1915–1987), who used to work at a Burns Hospital in London during the Second World War. He was treating patients injured and badly burned during bombing raids by the Germans. It was already known for some time that the best way to treat severe burns was to have it covered as soon as possible with healthy skin, preventing infection and loss of fluids. Ideally, the skin should be from the same patient, which then attached itself to the burned area and allowed healing to gradually occur. However, this was not always possible if the burnt area was extensive, in which case, thin pieces of donor's skin was used. This offered temporary relief as it would eventually slough off.

HLA Typing

HLAs (*Human Leucocyte Antigens*) are surface antigens found on all nucleated cells in the body. Every individual has a unique set of HLA antigens, inherited from each parent, which enables the body to differentiate "self" from "non-self" (foreign substances). Normally, all "non-self" tissues are rejected and rejection is faster if the difference between "self" and "non-self" is greater. It is for this reason that HLA typing is a pre-requisite for organ transplantation.

Three techniques are available for HLA typing. The first is the serological cytotoxicity method in which a small number of lymphocytes form blood are added to Terasaki plates. These plates have individual wells containing different specific antibodies. If there is binding between cells and antibodies the cells get killed in presence of complement. This enables differentiation of cells into various HLA types.

The second method uses flow cytometry. In this fresh leucocytes are added to monoclonal antibodies labeled with a fluorescent dye. Cells which bind to the antibody Fluoresce and are detected as they pass through a laser beam in the cytometer. This procedure takes about 30 minutes to complete.

The third method involves extracting DNA from cells and amplifying the genes that encode for HLA using PCR (polymerase chain reaction). The genes are the then matched with HLA sequences stored in gene bank. This method is particularly useful where precise matching is required, as in case of bone marrow transplants. This method may take several days to complete.

In some cases, due to the severity of the burns, a second transplant had to be done. Medawar noticed that if the *second* graft was taken from the *same* donor, the graft was rejected within a few days, while the first graft lasted for up to two weeks. Medawar rightly concluded that the earlier rejection of the second graft was due to immune reaction, engineered by the memory cells of the immune system. Medawar confirmed this hypothesis from experiments on mice, and for his work in immunology, he was awarded the Nobel Prize in 1960.

Subsequently it was shown that the rejection was caused by T cells which recognize the tissue antigens of the graft as "foreign" and immediately react against it. The T cells that cause the rejection are the cytotoxic T lymphocytes or CD8 "killer" T cells. These are different from CD4 "helper" T cells which are attacked by HIV virus.

The main strategy used at present to combat rejection is to administer immunosuppressants, which are often used in combination to decrease toxicity from any single agent. Most centers use triple therapies comprising Cyclosporin, Prednisolone and Azathioprine. A range of antisera against lymphocytes are also available which selectively inhibit their activity.

Although transplants are very useful, a number of life threatening complications can occur. Firstly, it could be due to the surgical procedure itself and secondly it could be due to infection, since transplant patients are extremely susceptible to a variety of infectious agents caused by immunosuppression.

The transplant patients also show a greater susceptibility for certain cancers such as skin cancers, blood cancers and a condition known as post-transplant lymphoproliferative disorders (PTLD), in which B cells proliferate, involving the central nervous system.

The graft–versus–host disease (GVHD) represents an exceptional situation, in which the donor T cells react against the *host* tissues. This occurs mainly in bone-marrow and bowel transplantation as these tissues carry many lymphocytes. In GVHD the patient develops rash, hepatitis and enteritis. In its chronic form, skin and joint degeneration may occur.

The kidney is the most commonly transplanted organ and is the preferred therapy for end stage renal disease. There are currently three possible sources of kidney donors. These are cadaver, living related, and living unrelated. The cadaver donation can be from a person who is brain dead or from a non-heart-beating cadaver.

For transplantation to succeed, the kidney should be in as good a physiological state as possible.

If the kidney comes from a living donor it is important to protect his interest. Informed consent is obviously needed and the donor should be told that the procedure is not completely risk free, with a donor mortality rate between 0.01 to 0.03% even in the best centers.

Corneal transplants differ from all other transplants, as the cornea has very limited blood supply which greatly reduces the possibility of rejection. Most corneal transplants work very well and may last for many years, without any need for immunosuppression.

Corneal transplants are recommended for patients who have suffered severe infection or injury of cornea; for inherited cornea thinning and inherited corneal clouding (Fuch's dystrophy). Corneal transplants are probably one of the most rewarding of operations, as they have contributed enormously to the quality of life of blind people all over the world.

TERMS USED IN TRANSPLANTATION

Graft:
The organ to be transplanted.

Autograft:
When the recipient and donor is the same individual. For example when skin is removed from one part of the body and applied to another part.

Isograft:
When the graft is between identical twins.

Allograft:
When the graft is between individuals of the same species.

Xenograft:
When the graft is between individuals of different species.

Orthotopic Transplant:
When the graft is placed in the same position as the recipient's diseased organ after it is removed e.g. heart and liver.

Heterotopic Transplant:
When the graft is placed in a different position. This is done when there is no advantage is removing recipients diseased organ from its normal location e.g. pancreas.

＊＊＊＊＊

"Medicine is a science of uncertainty and an art of probability".

Sir William Osler
(One of the best known English Physician)

23

Heredity and Disease

Heredity, or the transmission of traits from one generation to another, has puzzled mankind for centuries. It is obvious that semen is involved, but how does it interact in the female to produce an offspring which carries the parental characters?

The Greeks put forward many theories to explain this phenomenon. The Hippocratic school (500–400 BC) believed that semen is formed in different parts of the body before being transferred to testis. It carries the 'humors' emanating from each organ and these carry the hereditary traits. They believed that the female simply acted as an incubator for the baby to develop.

Aristotle (348–322 BC) was particularly interested in reproduction and marveled at the multiplicity of ways by which this goal was achieved in the animal kingdom. According to him, the female element contributes to the embryo's material and food, while the male element contributes to its energy and movement. The organs and characteristics of the embryo are formed by tiny particles which pass from every part of the adult into the reproductive elements. He disagreed with the Hippocratic school that all the parts of the embryo exists in a miniature form from the very beginning. He suggested that the parts develop under the shaping power of the "vital heat". As an example of embryonic development he describes how the chick embryo is formed, "…the embryo becomes first visible after three days…. The heart appears like a speck of blood, beating and moving as though endowed with life and from it, two veins with blood in them pass in a convoluted course and a membrane carrying bloody fibres from the vein ducts now envelope the yolk.... When the egg is ten days old, the chick and all its parts are distinctly visible". Aristotle implied that like the chick embryo, the human embryo also developed from stage to stage.

Another notion which had prevailed since ancient times was the fixity of species i.e. their form and shape never changed from the time of their appearance on Earth. The German botanist Gottleib Kolreuter (1733–1806) disproved this by producing hybrids of tobacco plant by cross breeding. This was a forerunner to a much more elaborate and famous work by Gregor Mendel (1822–1884), which forms the backbone of modern genetics.

Mendel was Abbot of Brno Monastery in Czech Republic, when he started his famous experiments on pea plants. He selected pea plants because they were easy to cross-breed and quick to grow. There were a number of pea plants: some were tall and some were short; some had smooth skin and some were rough; some produced white flowers and others purple. These were easily observable features and could be recorded after cross-breeding.

In his early experiments Mendel crossed pure tall with pure short, expecting that medium size plants would result. He was surprised to see that only tall plants appeared in the first generation. He obtained similar results with other traits i.e. only one trait appeared in the offsprings. He therefore concluded that some traits are *dominant,* while other traits which do not manifest are *recessive.* In another experiment, he took the plants resulting from crossing the tall with the short (impure) and crossed them with each other. He had already established that tallness was dominant so he expected all the offspring to be tall. However, the results were again surprising. In the subsequent generation the recessive trait had returned but only to one quarter of the plants. This happened because there were only four possible combinations for what Mendel called "units of inheritance", assuming that each plant has two units of opposing traits. If capital letter "A" is given to the dominant and small letter "a" to the recessive trait, the four possible combinations are AA, aA, Aa

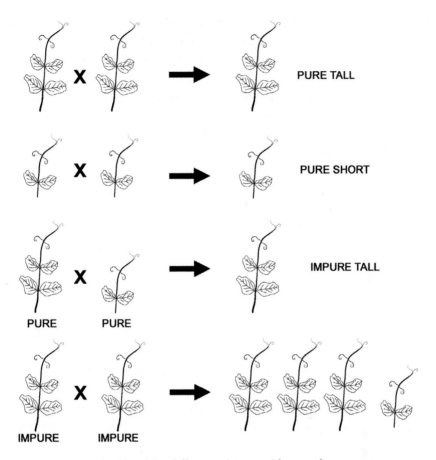

Fig. 23.1. Mendel's experiment with pea plants.

and aa. As the tallness is dominant and present in the first three combinations, all the three are tall. In the last combination (aa) the dominant trait (A) is absent, therefore, the recessive trait i.e. shortness appears (Fig. 23.1).

After Mendel, another discovery was made of incomplete or partial dominance. In this case, as neither characters are dominant, the offspring may appear with a mixed character. For example a plant called snapdragon having red flowers, is crossed with snapdragon of white flowers; the offspring may have pink flowers, as neither color is a dominant trait.

Mendel's abstract reasoning was brilliant and we now know that his units of inheritance are genes. A single gene carries instructions for building a single protein and gives our body the characteristics it possesses. As we inherit our genes from our parents, we look and perhaps even behave like them. This means we also inherit their diseases.

Actually, there is no disease in which genes do not have a role. The outcome and development of all diseases, to a certain extent, are determined by our genes. For diseases with clear exogenous causes such as infection, the host response is also influenced by the genes.

Genetically, determined diseases are usually classified into three categories: chromosomal disorders, single gene disorders, and polygenic disorders.

1. **Chromosomal Disorders:** These disorders occur as a result of deletion or addition of an entire chromosome or parts of chromosomes. As each chromosome contains a large amount of genes, its disorder manifests quite strongly and early in life. The loss, gain or major damage to a chromosome (other than sex chromosome) is usually incompatible with life and this often leads to abortion or miscarriage. It is estimated that 15% of all pregnancies are spontaneously aborted. Nature does not protect a grossly abnormal embryo or fetus.

A well known example of chromosomal disease is Down's syndrome in which there are 3 copies (trisomy) of chromosome 21 instead of 2 copies (Figs. 23.2 and 23.3). This abnormality occurs in approximately 1 in 800 live births and significantly increases with the advancing maternal age. Down's syndrome is characterized by mental retardation to a variable degree, growth retardation, congenital heart defects and physical deformity. Amongst the physical deformities are the upward-slanting eyes, giving the impression of mongoloid features, hence the term "mongolism" commonly used for the condition. Down's syndrome was the first genetic disease, in which the chromosomal deformity was recognized. It could be diagnosed, in the prenatal stage, by aminocentesis.

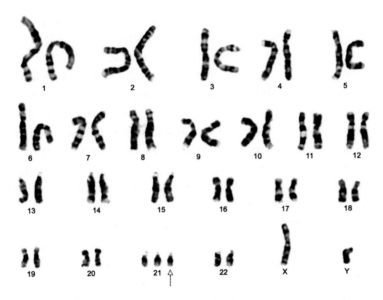

Fig. 23.2. Chromosome pattern in Down's syndrome in a male patient. Showing trisomy of chromosome 21. (Courtesy of AKU)

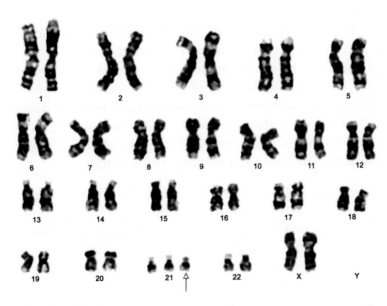

Fig. 23.3. Chromosome pattern in Down's syndrome in a female patient. Showing trisomy of chromosome 21. (Courtesy of AKU)

Defective genes may also be located on X and Y chromosomes when they get labeled as "sex-linked" genetic disorder. If a woman has a defective gene in one of her X chromosome, the normal X will produce the needed protein and she will not suffer from any disease. In contrast to this, if a male has a defect on his X-chromosome, his Y chromosome cannot compensate for this defect and he would suffer from the disease. The woman in this case becomes a carrier and will pass on the defective gene. Hemophilia is a classic example of a "sex-linked" genetic disease.

Hemophilia is also called the "royal disease" because Queen Victoria was a carrier and her son Prince Leopold suffered from it. Her daughters who got married to other European royal families spread it to them. The most notable case was Alexis, the only son of Czar Nicholas II of Russia who suffered from it.

Other examples involving X and Y chromosomes are Turner's syndrome in which there is only one sex chromosome or X chromosome monosomy (XO). In three other syndromes there is an extra sex chromosome which result in, XXY (Klinefelter's syndrome), XYY and XXX.

1. XO. This has a frequency of 1 in 2500 female births and results in gonadal hypoplasia and growth retardation.
2. XXY. This is the most common cause of male hypogonadism.
3. XYY. This is a rare disorder of males. Affected individuals are tall and may have behavioral problems and learning disability.
4. XXX. This has a frequency of 1 in 1000 female births. There is no distinguishing phenotype during infancy. They grow unusually tall with high incidence of infertility.

2. **Single Gene Disorders:** In these conditions only a single gene show mutation, but this may be enough to produce a profound effect on the person's health. These disorders are inherited in the Mendelian fashion showing autosomal dominant and recessive patterns. It is estimated that 5–10% of childhood mortality is because of these disorders. Amongst these conditions, familial hypercholesterolemia accounts for premature heart disease 1 in 500 individuals. Cystic fibrosis occurs 1 in 2000 whites in the USA. Sickle cell anemia affects 1 in 400 blacks in the USA. It is thought

that the high frequency of sickle cell gene occurs in people of African origin because carriers are more resistant than normal individuals to *Plasmodium falciparum* malaria. The sickle cell prevalence is a good example of Darwinian natural section as sickle cell was a lesser threat to survival than malaria.

The thalassemias are very important single-gene disorders involving hemoglobin synthesis. The disorder occurs in a broad belt of population from the Mediterranean region across the Middle East and the Indian sub-continent to South-East Asia. In some population groups, the carrier rate may be as high as 30%. Thalassemia major is a very serious disease, since the patients' survival depends on blood transfusion.

3. **Polygenic Disorders:** These diseases result from the interaction of multiple genes and are not well understood. They form the largest group of genetically determined conditions and include diabetes mellitus, schizophrenia, hypertension and most congenital heart diseases. Facial defects such as cleft lip and cleft palate fall under this group. These diseases account for approximately 25–35% of childhood mortality. The disease burden extends to the adult population because of their chronicity. The genetic hetrogenecity makes the rectification in these conditions difficult at the molecular level, even if such techniques become available.

GENETIC COUNSELLING[a]

Genetic diseases are at best preventable, as there is no reliable technology available at present to cure them, although the situation may change in years to come as great deal of research is going on in gene therapy, in which genes are intentionally introduced into human cells to effect a cure. For prevention, genetic counselling is thus very useful. During counselling, family history is taken and medical records checked. Recommendation is made for genetic tests and their results explained. The best

[a]Detail guidelines on genetic counselling is available on a website "March of Dimes". (http://www.modimes.org)

time to seek counselling is before pregnancy, when risk factors for the baby can be assessed.

The following risk factors exist:-

1. If either parent has an inherited disease.
2. If either parent already has children with genetic disease.
3. If the mother-to-be had repeated miscarriages or babies that died in infancy.
4. If the mother-to-be will be 35 or older when baby is born.
5. If a standard screening test e.g. alpha fetoprotein test yields an abnormal result.
6. If amniocentesis yields an abnormal result, e.g. Chromosomal defect.

Terms Used in Heredity

1. Gene — A hereditary unit that is transmitted from one generation to the other.

2. Genotype — The total genetic information present in an organism.

3. Phenotype — Is the external manifestation of the genotype — for example, the color determined by the hair color gene.

4. Chromosome — Structure residing inside the nucleus of a cell. It is made up of DNA and protein. Each species has a fixed number of chromosomes.

5. Mutation — The process that produces an alteration in DNA sequence of the gene.

6. Allele — *One* of any pair of alternate hereditary characters e.g. white or red eyes in *Drosophila* (fruit fly).

7. Heterozygote — An individual with different alleles at one or more loci. They will produce different gametes and therefore not breed true.

8. Homozygote — An individual with identical alleles at one or more loci. They will produce identical gametes and therefore breed true.

9. Hybrid — An individual produced by crossing parents of different genotypes.

*****

*"Variability is the law of life, and
as no two faces are the same, so no
two bodies are alike, and no two
individuals react alike and behave
alike under the abnormal conditions
which we know as disease".*

Sir William Osler
(One of the best known English Physician)

24

The End Game

The two most important events in life are birth and death, and one cannot exist without the other. It has been estimated that if *Staphylococcus* (bacteria) that exist on earth were allowed to multiply under optimal conditions, without any cell death, they would cover the entire surface of the earth 7 feet deep in 48 hours! This may be an exaggeration but it shows how necessary death is for sustaining life. When an organism dies, it replenishes the earth and new life forms arise. This is true, as much for bacteria, as it is for human being. In fact, all the activities in the universe are of a cyclical nature. Keeping this in mind, death is not the end but the beginning of life in another form (Fig. 24.1).

The study of death, at the cellular level, gives us an idea of its mechanism, and how this leads to the death of the organism itself. In the past, cell death was considered a passive process, but in 1972, this concept was challenged by a group of scientists from the University of Aberdeen, Scotland. They put forward the view that cells actually activate a program which results in their death. In other words, death is a purposeful activity of the cell.

Recent research confirms that cell death is genetically controlled and the name given to this suicidal act is apoptosis, from the Greek word meaning "falling off", as the petals fall off from the flowers. It is a universal phenomenon, and therefore a highly conserved feature of the evolutionary process.

Cyclical nature
of life and death*

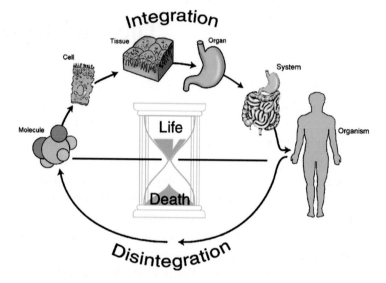

Fig. 24.1. The integration from the molecules to the organism is under the control of genes; the disintegration of tissues, after death, is due to the action of microbes in the soil. The elements released in the soil are utilized by plants. The plants are eaten by animals and this re-cycling process is continuously repeated in nature.

*Death refers here to the physical body. The existence of the spiritual body is a matter of personal belief.

Apoptosis is seen in the following situations:

1. The elimination of superfluous cells during embryonic development. For example, cells which form the interdigital web between the fingers and the toes, are removed by apoptosis after 6 weeks of gestation. Similarly, apoptosis also removes a large number of neuronal cells during the development of the brain.
2. The elimination of harmful cells. For example, virus-infected cells. In this case apoptosis is helped by T lymphocytes.

3. The elimination of cells which have undergone mutation, or whose DNA is damaged for any reason. For example, cancer cells.
4. The elimination of over produced germ cells. For example, oocytes (precursor of ova), are produced in many thousands, but only a maximum of 400 ova are extruded in the life-time of a female.

Apoptosis is a visible phenomenon and can be seen under the microscope. It is characterized by membrane blebing, condensation of chromatin, cell shrinkage and fragmentation of DNA, followed by absorption of the dead cell remnants by the neighboring cells. There is no inflammatory response by the body to apoptosis. This is in contrast to cell-death caused by artificial means, such as injury, in which inflammation occurs. It is estimated that about 10 billion cells die every day by apoptosis, and their place is taken by new cells, thus maintaining homeostasis or normal functioning of the body. However, not all populations of cells in the body are subject to turnover. Some cells formed during embryonic stage are retained throughout adult life, appearing not to divide and are irreplaceable. In mammals, cells of the central nervous system, muscle cells of the heart, lens cells of the eye and auditory hair cells of the ear fall into this category.

It has been said that "Man in an obligate aerobe", which in simple language means we cannot live without oxygen. If the amount of oxygen reaching our tissues falls below a certain level, cells start dying. If this process continues, cessation of all the vital functions occurs; the heart beat, the brain activity and breathing cease resulting in death. However, *brain death* can occur on its own with other organs sill functioning or made to function by artificial means. Brain death is defined as "irreversible cessation of all functions of the entire brain, including the brainstem". Death can come suddenly as in accidents or more gradually as in ageing due to cumulative degenerative processes.

Why does ageing occur? There are a number of theories to account for this. One of this relates to telomeres, which are structurally located on the tips of chromosomes. With age, the telomeres get shorter and are unable to maintain the structural integrity of the chromosomes. Telomeres themselves are maintained by the enzyme telomerase, which is available only in a limited quantity in the cell. As a result, every time a cell divides, the telomere length decreases. This process acts as an internal clock which, after a period of time, causes the cell to stop dividing and apoptosis to ensue. A recent study suggests that it may be possible to delay senescence by artificially increasing the level of telomerase.

It has been known since the 1950s that free radicals (oxygen molecules generated during metabolism) can damage cells including their DNA and could be an important cause for ageing. It has therefore been suggested that taking antioxidants such as Vitamins C and E may delay the ageing process. However, there is no firm evidence to support this.

Ageing affects every organ of the body, as they are all made up of cells. The normal defenses that protect us from microbes grow weaker, resulting in a rapid deterioration, as in the case of an infection, such as pneumonia. There is also a decline in the lean body mass and a corresponding increase in fat, affecting the distribution of drugs in the body; therefore, medication for the elderly needs to be reduced. Meanwhile, fibrous tissue in the heart and blood vessels increases, making them less elastic. Owing to this, the heart is slower in responding to stress and to physical exertion.

While the lungs, kidneys and liver functions deteriorate only to a minor degree, gastric or stomach cells usually atrophy faster and may cause vitamin B_{12} deficiency. As the bones begin to weaken, the chances of fractures increase and the joints too show signs of wear and tear. Physical changes in the eyes and ears are also common. A decrease in the elasticity of the lens causes impairment of near vision, while degeneration of hair cells (Fig. 24.2) on the organ of Corti[a] in internal ear result in high-tone deafness.

The hormonal system undergoes a transformation that manifests itself in a decrease in sexual activity in men and menopause in women. Men are generally afflicted with urinary

---

[a]An Italian Anatomist who first described this structure.

Fig. 24.2. V-shaped rows of hair cells on the organ of Corti in the internal ear. They carry sound waves along the nerve pathways to the brain. ×4300. SEM. (Courtesy of Ms. May Lee Mui Gek and Dr. Runsheng Ruan, Institute of Bioengineering and Nanotechnology, Singapore)

problems as well, often caused by the enlargement of the prostate gland located at the neck of the urinary bladder.

There is some interesting data on the effect of reproduction on ageing, using the fruit fly, *Drosophila melanogaster*. Scientists at the University College, London, bred various lines of fruit flies; some of which laid eggs at a very young age, whereas others followed suit only when they were older. It was found that the flies that bred young, died earlier than the "older" ones. However, when the young reproducing flies were sterilized using X-rays, their life span also became equally long. This led to the conclusion that egg production reduced longevity in the female fruit flies. In theory, this implies that once an animal has reproduced — passed its deoxyribonucleic acid or DNA to its offspring — the ageing process, which ultimately leads to death, sets in. "The reason why you and I will grow old and die, I am sorry to say, is that we become dispensable", writes British gerontologist Tom Kirkwood in his book, *Time of Our Lives*.

Recent experiments carried out at the European Institute of Oncology reveal that targetted deletion of a protein known as p66shc, not only increased the life span of mice by one-third, but also resulted in their remaining healthy during this extended period. Similarly, in *Drosophila* two genetic mutants, known as "Mathusela" and "Indy" (for I'm not dead yet) increased their life span by over 30%. The human genome is clearly more complex than that of mice and *Drosophila*, but the laws of biology apply uniformly to all life forms. Therefore, it might just be possible that an analogous genetic change in the human genome could turn us into centenarians!

The genetic diagnosis of diseases is already available to us. This data is bound to increase in the future, helping people to discover their natural chances of living up to a certain age, as well as the possibilities of contracting certain diseases. This will enable us to take preventive measures before the disease manifests. A good example of this is diabetes, which if not diagnosed and treated early, can cause blindness, strokes and heart attacks.

Just as an individual's genetic make-up plays a significant role in determining our life span, so does the environment, since we are a product of both nature and nurture. The ingestion of chemical toxins in water and food, cosmic, ultraviolet and X-rays, and even gravity, are some of the environmental factors that may affect longevity. Meanwhile, a healthy lifestyle based on regular exercise, sensible eating habits and effective stress management is also important. Exercise retards age-related bone loss and improves cardiac performance. Research in the US has shown that mail delivery staff who walk a lot, on a average live two years longer than mail sorters and other sedentary postal employees.

It is important, in fact crucial, to opt for a form of activity that one is comfortable with. An aged person may not be able to indulge in strenuous exercise. Indeed, it may be dangerous to do so. Yoga is ideally suited for the aged, as it does not tire the person and is good for the aches and pains synonymous with old age. This activity also allows people to become more sensitive to their body language, as they realize the strengths or weaknesses of various joints and limbs by performing positions known as *mudras*. Additionally, yoga tones up the internal organs such as the stomach and intestines.

Over-indulgence in food high in animal fat and calories, predisposes the body to obesity, diabetes and heart disease, which in turn, leads to further health problems. Fast food chains which have sprung up all over the world are at least partly responsible for this. In mice calorie restriction increases their life span. Smoking is another important factor behind a multitude of illnesses, since cigarettes contain chemicals that damage the DNA and cause cancer. While these physical factors have a direct impact on how a person ages, mental well-being is no less crucial. Senior citizens are generally more prone to psychological stress related to retirement, health issues, financial difficulties and loneliness. So it is all the more vital for the aged to understand the process they are undergoing. Ageing is a gradual process. It is more like going down step by step, rather than rapidly moving down an escalator. The records of marathon runners show that some could still run in their 70s or 80s, even though the time required for completion was much longer. This decrease in performance occurred gradually over many decades. In other words homeostatic (adjustment) mechanism gets impaired with ageing and reversion to normality takes longer.

Remaining mentally active is very good for the nervous system, which is not very different from the muscles of the body. Similar to unused muscles, which grow weak over a period of time — a phenomenon known as disuse atrophy — nerve cells too require constant stimulation in order to remain functional. There is, for instance, enough evidence to suggest that higher education is associated with a lower risk of Alzheimer's disease.

How we view old age also has an impact on our health and behavior. This image is dependent on various factors that include both cultural and social assumptions, as well as individual experiences. Ellen Langer and her colleagues at Harvard University, examined the effect of a person's childhood experience with their grandparents, on their own ageing. It was found that, the images of old age that they picked up from their grandparents, was important. If the grandparents were active when the individual first got to know them, they developed a more "youthful" image of old age, thus approaching their own ageing process more positively.

On reaching the age of retirement, we need to decide how to manage our lives. Some of us take the path of seclusion and find comfort in spiritual activities, experiencing what is being referred to as "growth through diminishment". The process that a seed undergoes, on being planted in the ground, provides a befitting analogy. As the seed decays, it sprouts new life from the nurturing forces of the earth.

The other path is that of unceasing activity and continued involvement with work. Bertrand Russell, the great secular philosopher, was a strong believer in this approach and campaigned for nuclear disarmament well into his 90s. Just before his death, he talked of his passions: "the longing for love, the search for knowledge and unbearable pity for the suffering mankind".

Another advocate of remaining active throughout one's life, Robert Louis Stevenson, wrote: "By all means begin your folio; even if the doctor does not give you a year, even if he hesitates about a month, make one brave push and see what can be accomplished in a week. It is not only in finished undertakings that we ought to honor useful labor. All who have meant good work with their whole hearts have done good work, although they may die before they have time to sign it".

Indeed, old age will never be completely conquered, as death must come to all. But if we are lucky enough to inherit good genes and watch our health, there is no reason why the end game should be unmanageable.

Ending on a philosophical note, I concur with John McManners, as quoted by D.J. Enright in his book *Death*. He wrote, "The attitude of men to death of their fellows is of unique significance for an understanding of our human condition... The knowledge that we must die gives us our perspective for living, our sense of finitude, our conviction of the value of every moment, our determination to live in such a fashion that we transcend our tragic limitation".

Hayflick Limit

Leonard Hayflick, Professor of Anatomy, University of California is well known for his experiments which show that there is a limit to the replicative capacity of normal cells. The limit could be demonstrated in cells grown *in vitro*. For example, when fibroblasts are removed from the umbilical cord of a fetus and grown in a culture medium, they divide until they are dense enough to come in contact with each other, then they stop dividing due to a signaling mechanism, know as contact inhibition. However, if these cells are diluted and cultured again they start dividing again. This process can be repeated approximately 50 times after which they permanently *cease* to divide. The upper limit of cell division varies from animal to animal depending on their life span. For example, animals with a short life span, such as mice, undergo 10–15 divisions. In contrast to this, Tortoises which have long life span (about two hundred years) can undergo over 100 divisions. Due to this biological limitation, Hayflick estimates that the upper limit of human life is about 125 years.

✳✳✳✳✳

"I depart as air, I shake my white locks at the runaway sun;
I effuse my flesh in eddies, and drift it in lacy jags.
I bequeath myself to the dirt, to grow from the grass I love;
If you want me again, look for me under your boot-soles.

You will hardly know who I am, or what I mean;
But I shall be good health to you nevertheless,
And filter and fiber your blood.

Failing to fetch me at first, keep encouraged;
Missing me one place, search another,
I stop somewhere, waiting for you".

Walt Whitman
Song of Myself

Hayflick Limit

Leonard Hayflick, Professor of Anatomy, University of California, is well known for his experiments about how their lives as a limit on the reproductive capacities of animal cells. The limit could be illustrated in cell-growth in vitro. For example, when fibroblasts are extracted from the umbilical cord of a fetus and grown in a culture medium, they divide until they are dense enough to come in contact with each other, then they stop dividing due to a signaling mechanism known as contact inhibition. However, if these cells are diluted and culture serum they start dividing again. This process can be repeated approximately 50 times, after which they reach the Hayflick limit. The average lifespan of cell division varies from animal to animal, depending on their life span. For example, animals with a shorter life span, can be quite short, reaching 10–15 divisions. In contrast to this, tortoises, which have a life span almost two hundred and fifty corresponding 120 divisions. Here the biological limitation for the human body is that the total lifetime for human life is about 120 days.

Annex

HOW LONG WILL YOU LIVE?

The life insurance companies have devised a questionnaire to estimate life expectancy. It is only a *guide* and its accuracy and reliability has not been critically assessed.

Begin with the number 76, then add or subtract the value of (– means subtract, + means add). The final number on your total will provide a rough estimate of your life expectancy.

Personal Facts	Calculation	Subtotal
If you are male	−3	_____
If female	+4	_____
If you live in an urban area with a population over 2 million	−2	_____
If you live in a town under 10000 or on a farm	+2	_____
If a grandparent lived to 85	+2	_____
If all four grandparents lived to 80	+6	_____
If either parent died of stroke or heart attack before the age of 50	−4	_____
If any parent, brother or sister under 50 has (or had) cancer or a heart condition, or has had diabetes since childhood	−3	_____
Do you earn over $50,000 a year?	−2	_____
If you finished college	+1	_____
If you have a graduate or professional degree	+2	_____
If you are 65 or over and still working	+3	_____
If you live with spouse or a friend	+5	_____
If you do not live with a spouse or a friend	−3	_____
Subtract 3 for every decade lived alone since age 25		

Age Adjustment		
If you are between 30 and 40	+2	_____
If you are between 40 and 50	+2	_____
If you are between 50 and 70	+4	_____

Personal Facts	Calculation	Subtotal
Lifestyle Status		
If you work behind a desk	−3	_____
If your work requires regular, heavy physical labor	+3	_____
If you exercise strenuously (tennis, running, swimming etc.)		
Five times a week for around half an hour	+4	_____
Two or three times a week	+2	_____
Do you sleep more than ten hours each night?	−4	_____
Are you intense, aggressive, easily angered?	−3	_____
Are you easy going and relaxed?	+3	_____
Are you happy?	+1	_____
Are you unhappy?	−2	_____
Have you had a speeding ticket in the past year?	−1	_____
Do you smoke more than two packs a day?	−8	_____
One or two packs?	−4	_____
One-half to one packs?	−3	_____
Do you drink the equivalent of 1 oz. of liquor a day?	−1	_____
Are you overweight by 50 lbs. Or more?	−8	_____
By 30 to 50 lbs?	−4	_____
By 10 to 30 lbs?	−2	_____
If you are a man over 40 and have annual checkups	+2	_____
If you are a woman and see gynecologist once a year	+2	_____
	Grand Total	_____

http://www.gfcwow.com/download.htm

Recommended Books for Further Reading

This is a personal selection of a few non-technical books from the numerous that are available in life sciences. I have provided a brief comment on each to act as a guide to their contents. Almost all of them are short and easy to read and I concur with *Ecclesiastes* 12:12 that, "much study is weariness of the flesh".

1. *The Ascent of Science*

By   Brian L. Silver
Publ: Oxford University Press, 1998

An excellent introduction to general science. Covers biology, physics and chemistry and gives historical background to many important discoveries and inventions. This book is a fitting successor to Bronowski's classic, *The Ascent of Man*.

2. *Life on Earth*

By   David Attenborough
Publ: William Collins, 1979

This book is a gold-mine of information on the history of life on Earth. It has stunning photographs of unmatched beauty, although published more than 20 years ago. An equally good more recent publication is *The Private Life of Plants*. I don't think any one has done more to popularize life sciences than David Attenborough through his publications and TV programmes.

3. *The Blind Watchmaker*

By   Richard Dawkins
Publ: Penguin books, 1986

The author presents a forceful defense of the Darwinian evolution. He thinks that Darwinism applies not only to this planet, but all over the universe, wherever life may be found. Dawkins' earlier book, *The Selfish Gene* (1976) is also very thought provoking. He asserts that the living world is nothing more than DNA replicating machine but "we have the power to defy the selfish genes of our birth". In other words our innate selfishness can be controlled by our brain power. Where this brain power comes from is not answered.

4. *Origins*

By   Hubert Reeves, Joel De Rosney,
          Yves Coppens and Dominique Simonet
Publ: Arcade Publishing, 1998

Three eminent scientists — an astrophysicist, an organic chemist and an anthropologist, discuss the origin of the universe and life on earth. Important questions about science and religion are also covered. Science learns whereas religion teaches. Doubt is the motivation force behind the former; whereas faith glues the latter.

5. *The Astonishing Hypothesis*

By   Francis Crick
Publ: Simon and Schuster, 1995

The book is the scientific search for the soul. It tackles the crucial issue of our existence i.e. the nature of consciousness. It is a good book on how the brain functions. Crick points out that the brain not only formulates words to denote objects but it can also develop abstract concepts. This ability could lead to self-deception, making it almost inevitable for us to jump to the wrong conclusions, especially about rather abstract matters, without the discipline of scientific research.

6. *Lamarck's Signature*

By   Edward J. Steele, Robyn A. Lindley and
        Robert V. Blanden
Publ: Perseus Books, 1998

This is amongst the few publications giving support to Lamarck's ideas on evolution. The authors, in my opinion, correctly argues that *random* genetic mutation are probably not the *only* mechanistic agent of evolutionary change. The prevailing dogma that in a multicellular organism somatic modifications cannot be inherited (Weismann's barrier), is refuted by the fact that aspects of acquired immunity developed by parents, in their lifetime, can be passed to their offsprings.

7. *The Malaria Capers*

By   Robert S. Desowitz
Publ: Norton, 1993

The author calls the book "tales of parasites and peoples". The book deals mainly with two major medically important parasites, *Leishmania donovani* and *Plasmodium* spp. The former causes Kala-Azar (visceral leishmaniasis) and the later Malaria. The book is as interesting as any novel, with leading characters coming and going, as actors in a play. It reveals how scientific discoveries are made by dedication, hardwork and sometimes shere good-luck. Desowitz makes an important observation that the traditional parasitologists, who are trained to recognize parasites, are being rapidly replaced by molecular biologists, which in the long run could be detrimental to the study of tropical diseases.

8. *How We Die*

By   Sherwin B. Nuland
Publ: Alfred A. Knopf, 1994

Death is of universal concern but not adequately covered even in medical text books. The author gives a graphic description of dying and how death comes in its various forms. It covers death due to Alzheimer's disease, accidents, suicide, stroke, heart attack, AIDS and cancer. He reveals that death is by no means as peaceful as is often described, but we need to know its social, moral and ethical implications.

9. *Rocks of Ages*

By   Stephen Jay Gould
Publ: The Baltimore Publishing, 1999

A brilliant discussion of the status of religion and science in society. Science tries to document factual characteristics of the natural world, while religion operates in the realm of meaning and values. He calls them the two "magisteria". They are separate and should remain separate.

Index